Recurrent Events Data Analysis for Product Repairs, Disease Recurrences, and Other Applications

ASA-SIAM Series on Statistics and Applied Probability

The ASA-SIAM Series on Statistics and Applied Probability is published jointly by the American Statistical Association and the Society for Industrial and Applied Mathematics. The series consists of a broad spectrum of books on topics in statistics and applied probability. The purpose of the series is to provide inexpensive, quality publications of interest to the intersecting membership of the two societies.

Editorial Board

Robert N. Rodriguez
SAS Institute Inc., Editor-in-Chief

Janet P. Buckingham
Southwest Research Institute

Richard K. Burdick
Arizona State University

James A. Calvin
Texas A&M University

Katherine Bennett Ensor
Rice University

Douglas M. Hawkins
University of Minnesota

Lisa LaVange
Inspire Pharmaceuticals, Inc.

Gary C. McDonald
National Institute of Statistical Sciences

Paula Roberson
University of Arkansas for Medical Sciences

Dale L. Zimmerman
University of Iowa

Ross, T. J., Booker, J. M., and Parkinson, W. J., eds., *Fuzzy Logic and Probability Applications: Bridging the Gap*

Nelson, W. B., *Recurrent Events Data Analysis for Product Repairs, Disease Recurrences, and Other Applications*

Mason, R. L. and Young, J. C., *Multivariate Statistical Process Control with Industrial Applications*

Smith, P. L., *A Primer for Sampling Solids, Liquids, and Gases: Based on the Seven Sampling Errors of Pierre Gy*

Meyer, M. A. and Booker, J. M., *Eliciting and Analyzing Expert Judgment: A Practical Guide*

Latouche, G. and Ramaswami, V., *Introduction to Matrix Analytic Methods in Stochastic Modeling*

Peck, R., Haugh, L., and Goodman, A., *Statistical Case Studies: A Collaboration Between Academe and Industry, Student Edition*

Peck, R., Haugh, L., and Goodman, A., *Statistical Case Studies: A Collaboration Between Academe and Industry*

Barlow, R., *Engineering Reliability*

Czitrom, V. and Spagon, P. D., *Statistical Case Studies for Industrial Process Improvement*

Recurrent Events Data Analysis for Product Repairs, Disease Recurrences, and Other Applications

Wayne B. Nelson

Schenectady, New York

siam
Society for Industrial and Applied Mathematics
Philadelphia, Pennsylvania

ASA
American Statistical Association
Alexandria, Virginia

Copyright © 2003 by the American Statistical Association and the Society for Industrial and Applied Mathematics.

10 9 8 7 6 5 4 3 2 1

All rights reserved. Printed in the United States of America. No part of this book may be reproduced, stored, or transmitted in any manner without the written permission of the publisher. For information, write to the Society for Industrial and Applied Mathematics, 3600 University City Science Center, Philadelphia, PA 19104-2688.

The correct bibliographic citation for this book is as follows: Nelson, Wayne B., *Recurrent Events Data Analysis for Product Repairs, Disease Recurrences, and Other Applications*, ASA-SIAM Series on Statistics and Applied Probability, SIAM, Philadelphia, ASA, Alexandria, VA, 2002.

Excel is a trademark of Microsoft Corporation in the United States and other countries.
SAS/QC and JMP are registered trademarks of SAS Institute, Inc.
S-PLUS is a trademark of Insightful Corporation.
Weibull++ is a trademark of Reliasoft Corporation.

Library of Congress Cataloging-in-Publication Data

Nelson, Wayne B., 1936-
 Recurrent events data analysis for product repairs, disease recurrences, and other applications / Wayne B. Nelson.
  p. cm. — (ASA-SIAM series on statistics and applied probability)
 Includes bibliographical references and index.
 ISBN 0-89871-522-9
1. Failure time data analysis. 2. Survival analysis (Biometry) I. Title. II. Series.

QA276 .R465 2002
519.5—dc21

2002030864

siam is a registered trademark.

*This book is gratefully dedicated
to my many generous clients, friends,
and colleagues who contributed to it.*

Contents

Preface   ix
     Acknowledgments   x

1   Recurrent Events Data and Applications   1
     1.1   Introduction   1
     1.2   Exact Ages and Right Censoring   3
     1.3   Exact Ages, Left Censoring, and Gaps   10
     1.4   Interval Age Data   11
     1.5   Continuous Histories   13
     1.6   A Mix of Types of Events   15
     1.7   Practical Issues   18
     Problems   21

2   Population Model, MCF, and Basic Concepts   23
     2.1   Introduction   23
     2.2   Cumulative History Functions   23
     2.3   Population Model and Its MCF   25
     2.4   Information Sought   28
     Problems   33

3   MCF Estimates for Exact Age Data   35
     3.1   Introduction   35
     3.2   MCF for Number from Exact Age Data with Right Censoring   37
     3.3   MCF for Cost from Exact Age Data with Right Censoring   43
     3.4   MCF from Exact Age Data with Left Censoring and Gaps   46
     3.5   MCF from Continuous History Function Data   49
     3.6   Practical and Theoretical Issues   51
     Problems   55

4   MCF Confidence Limits for Exact Age Data   59
     4.1   Introduction   59
     4.2   Nelson's Confidence Limits   59
     4.3   Naive Confidence Limits   64
     4.4   Assumptions and Theory   68
            4.4.1   The MCF Estimate $M^*(t)$ 68
            4.4.2   Variance of the Estimate and Confidence Limits   70
     Problems   76

5	**MCF Estimate and Limits for Interval Age Data**		**79**
	5.1 Introduction		79
	5.2 MCF Estimate		79
	5.3 Confidence Limits		83
	Problems		87
6	**Analysis of a Mix of Events**		**97**
	6.1 Introduction		97
	6.2 Model for a Mix of Events		98
	6.3 MCF with All Types of Events Combined		100
	6.4 MCF for a Single Type of Event		102
	6.5 MCF for a Group of Events		103
	6.6 MCF with Events Eliminated		105
	6.7 Practical and Theoretical Issues		107
	Problems		108
7	**Comparison of Samples**		**109**
	7.1 Introduction		109
	7.2 Pointwise Comparison of MCFs		110
	7.3 Comparison of Entire MCFs		116
	7.4 Theory, Issues, and Extensions		118
	Problems		118
8	**Survey of Related Topics**		**121**
	8.1 Introduction		121
	8.2 Poisson Process		121
	8.2.1 Poisson Process Model		122
	8.2.2 Single-Sample Analyses		124
	8.2.3 Multisample Analyses		126
	8.3 Nonhomogeneous Poisson Processes		127
	8.4 Renewal Processes		130
	8.5 Models with Covariates		132
	8.6 Other Models		133
	Problems		134
References			**137**
Index			**143**

Preface

Time-to-event data are important in many applied fields. Examples include time to failure of nonrepairable products, survival time of medical patients, length of time bank accounts stay open, length of time that subscribers pay for cable service, and time recipients spend on welfare. Methodology for such data is called reliability data analysis, life data analysis, survival analysis, and time-to-event analysis. Most books present models and data analysis methods for sample units that experience only one event, end of life. Then each sample unit has one observed value, its age at "death" or else its current age while still "alive." The population yielding such data is typically modeled with a simple life distribution, such as the Weibull or lognormal.

In contrast, many applications involve repeated events data where a sample unit may accumulate any number of events over time. Examples include the following:

- the number and costs of repairs on machines, cars, electronics, appliances, and other repairable engineered products;

- the number and treatment costs of recurrent disease episodes in patients (e.g., bladder tumors, strokes, and epileptic seizures);

- the number and costs of childbirths, divorces, reincarcerations, reincarnations, etc.

Such repeated events (or recurrence) data are modeled with a stochastic point process; the Poisson process is the simplest such model. Suitable nonparametric models and data analyses for recurrent events appear in few books; however, much of this literature is not written for practitioners who need such methodology.

Meant as an introduction, this book provides basic nonparametric methods for such data, especially the plot of the nonparametric estimate of the population mean cumulative function (MCF), which yields most of the information sought. This plot is as informative as is the probability plot for life and other univariate data. Also, this book is the first to present a simple unified theory that includes data on "costs" or other "values" of discrete events; in contrast, most previous work applies only to counts of event recurrences. For practitioners, this book surveys and gives output of computer programs that calculate and plot the MCF estimate with confidence limits. Many such calculations can easily be done with a pocket calculator or spreadsheet program. In addition, this book provides methods for accumulated variables that vary continuously with time; for example, the power output of a power plant is integrated over time to determine its total energy output. Also provided are selected references for those who seek further details on this and related subjects.

The intended audience for this book includes engineers (in reliability, design, and maintenance), medical researchers, social scientists, econometricians, marketing researchers, criminologists, statisticians, parents, divorcees, felons, and other practitioners who encounter repeated events data. The background needed for most of this book is a basic statistics course.

Previous knowledge of stochastic processes is not needed, as the simple presentation here is self-contained. Technical sections explain assumptions about the relevant theory, models, and methods. These assumptions must be verified in practice to ensure that results of the analyses are valid.

Most applications in this book come from the author's consulting experience with reliability data in industry. Other examples include recurrent bladder tumors and childbirths. However, workers in other fields will find that the models and methods here apply to most other kinds of repeated events data. Most data sets can be downloaded from http://www.siam.org/books/sa10/.

The book contains the following major topics/chapters:

1. *Recurrent Events Data and Applications*: This surveys types of recurrence data and information sought.

2. *Population Model, MCF, and Basic Concepts*: This presents the basic nonparametric population model, its mean cumulative function (MCF), and other concepts.

3. *MCF Estimates for Exact Age Data*: This shows how to estimate, plot, and interpret the sample MCF for exact age data, the most common form of age data.

4. *MCF Confidence Limits for Exact Age Data*: This gives confidence limits for the MCF and their interpretation.

5. MCF *Estimate and Limits for Interval Age Data*: This shows how to estimate, plot, and interpret the sample MCF for interval age data, a common form of age data.

6. *Analysis of a Mix of Events*: This shows how to calculate, plot, and interpret MCFs for data with a mix of types of events.

7. *Comparison of Samples*: This presents methods for comparing two or more samples of recurrent events data.

8. *Survey of Related Topics*: This surveys parametric and regression models for recurrence data and other related topics.

References.

Acknowledgments

This book has benefited from generous contributions of many friends and colleagues, many of whom have contributed to the methodology for the analysis of recurrence data. The author is also grateful to his clients who have generously permitted their data to appear here. These include Mr. Richard J. Rudy (car transmissions), Mr. Nance C. Lovvorn (fan motors), Mr. Jim Gay (turbines), Dr. David P. Ross (compressors), Mr. Virgil Wheaton (traction motors), Mr. Lorne MacMonagle (braking grids, valve seats), Mr. Charlie Bedford (circuit breakers), and Mr. Lee Morrison (defrost controls). Other generous clients prefer to remain unnamed. Computer output and plots were contributed by Dr. Gordon Johnston and Dr. John Sall of the SAS Institute from SAS PROC Reliability and JMP®; by Prof. William Q. Meeker from his SPLIDA routines for S-PLUS™; and by Mr. Pantelis Vassiliou and Mr. Adamantios Mettas of ReliaSoft Corporation from their Recurrence Data add-on to the Weibull++6 package. The manuscript has benefited from input, reviews, and suggestions from Dr. Dave Trindade, Prof. William Q. Meeker, Dr. Necip Doganaksoy, Mr. Harry Ascher,

Prof. Rita Aggarwala, Dr. Gordon Johnston, Mr. Pantelis Vassiliou, Dr. Bob Abernethy, Luís A. Escobar, Mr. Nick Zaino, Dr. Larry Crow, Dr. Jeff Robinson, Prof. Jerry Lawless, Ms. Anne Brown, Prof. Wally Blischke, Prof. Paul Allison, and Prof. Steve Rigdon. Mr. Ted Gardner generously provided much needed computer help and advice. Ms. Carolyn Micklas provided much appreciated help with certain word processing. The SIAM staff were also most helpful, particularly Ms. Alexa Epstein, Ms. Sara Murphy, the production staff, and Dr. Linda Thiel. Dr. Bob Rodriguez merits acknowledgment for his helpful personal attention and contributions as Editor-in-Chief of the ASA-SIAM series. The author is most grateful to the Fulbright Commission for a Fulbright Award for research and lecturing in Argentina, where the author did some final work on this book.

Wayne Nelson
Schenectady, NY
wnconsult@aol.com

February, 2002

Chapter 1
Recurrent Events Data and Applications

1.1 Introduction

Purpose. This chapter presents typical types of recurrent events data, also called repeated events data and recurrence data. Such data arise in many fields. For example, the applications here include transmission repairs in cars, recurrences of bladder tumors in a medical study, and births of children to statisticians. The chapter also presents basic concepts, using applications for motivation, describes information sought from such data, and discusses important practical issues. This chapter is essential background for later chapters. This section first surveys some fields where recurrent events data arise, next describes two approaches to modeling such data, and finally overviews the remainder of the book and chapter.

Applied fields. Recurrent events data arise in various applied fields. The following list gives some fields, references, and applications in this book.

- *Product reliability*: General reliability references on recurrence data include this book, Rigdon and Basu (2000), Meeker and Escobar (1998, Chapter 16), Ascher and Feingold (1984), Lawless (1983), and Ascher (2003). Industry applications include the following:

    - *Automotive*: Kalbfleisch, Lawless, and Robinson (1991), Nelson (1998, 2000b), Robinson (1995), Robinson and McDonald (1991), Hu, Lawless, and Suzuki (1998), Lawless, Hu, and Cao (1995), Lawless and Nadeau (1995), Suzuki (1985, 1993); car transmissions (section 1.2, Problem 3.1), gearboxes (Problems 5.6 and 8.4).

    - *Computers/electronics*: Tobias and Trindade (1995), Trindade and Haugh (1980), Vallarino (1988), Nelson (1988).

    - *Electric power*: Nelson (1990), Ross (1989), Kvam, Singh, and Whitaker (2002); heat pumps (section 1.2), power circuit breakers (Problems 3.5 and 4.5).

    - *Transportation*: Davis (1952), Doganaksoy and Nelson (1991, 1998), Nelson and Doganaksoy (1989); traction motors (section 1.6), valve seats (Problem 3.4), braking grids (section 7.2).

    - *Military*: Crow (1982), Duane (1964), IEC Standard 1164 (1995), MIL-HDBK-781 (1987), Ascher and Feingold (1984); turbines (section 2.4).

- *Appliances*: Nelson (1979), Agrawal and Doganaksoy (2001); compressors (section 3.4), defrost control (Problem 5.2).

- *Aviation*: Proschan (2000).

- *Medical equipment*: Baker (2001), Nelson (1988, 1995b); blood analyzers (section 2.4).

• *Medicine*: Byar (1980), Therneau and Hamilton (1997), Therneau and Grambsch (2000), Wang, Qin, and Chiang (2001), Thall and Lachin (1988), Fleming and Harrington (1991), Andersen et al. (1993), Peña, Strawderman, and Hollander (2001), Lawless (1995b), Cook, Lawless, and Nadeau (1996), Cameron and Trivedi (1998); bladder tumors (section 1.2); CGD (Problems 3.6 and 4.6).

• *Social sciences*: Allison (1984, 1996), Harris (1996), Kraatz and Zajac (1996), Myers (1997); childbirths (section 1.4).

• *Economics*: Lancaster (1990), Heckman and Singer (1985), Cameron and Trivedi (1998), Winkelmann (2000).

• *Business/marketing*: Winkelmann (2000), Morin (2002).

• *Criminology*: Cohen et al. (1998), Maltz (1984).

Times. Some previous work models and analyzes the times *to* first recurrence, times *to* second recurrence, etc., using distributions. Davis (1952) is an early example of this approach, and Wei, Lin, and Weissfeld (1989) is a more recent example. Another approach models and analyzes the times *between* recurrences. Most of this work applies to data on just a single unit, for example, Ascher and Feingold (1984), whose initial model is a renewal process. That is, the times between recurrences are independent observations from the same distribution. When this simple renewal model is inadequate, they characterize how the data depart from the model; for example, interarrival times are neither independent nor identically distributed. Use of times *to* and *between* recurrences is cumbersome and yields limited information only from counts of recurrences—not from costs or other values of recurrences.

Stochastic processes. Other work, including this book, uses counting process models and analyses for the *number* of recurrences. Tobias and Trindade (1995), Meeker and Escobar (1998), and Therneau and Grambsch (2000) provide basic nonparametric methods for counting data, but they do not show the full power and versatility of available methods. Aalen (1978), Fleming and Harrington (1991), and Andersen et al. (1993) employ martingale theory to obtain asymptotic results for counting data. Englehardt (1995), Rigdon and Basu (2000), and other authors use parametric counting process models and analyses. These often require dubious assumptions, such as the use of a nonhomogeneous Poisson process and a common process for all units. Most such literature applies only to *counts* of recurrences. In contrast, this book provides simpler and more general nonparametric methods that readily yield all information sought. These methods also apply to *cost* and other observed *values* of events, not just counts.

Book overview. This book contains the following chapters:

1. *Recurrent Events Data and Applications*: This presents various types of recurrence data, the information sought from such data, and basic concepts.

2. *Population Model, MCF, and Basic Concepts*: This presents the basic nonparametric population model, the mean cumulative function (MCF, which contains most information sought from data), basic concepts, and information sought from fitting the model to data.

3. *MCF Estimates for Exact Age Data*: This explains how to estimate, plot, and interpret the sample MCF to obtain the information sought from exact age data.

4. *MCF Confidence Limits for Exact Age Data*: This shows how to obtain and interpret confidence limits for the MCF for such data.

5. *MCF Estimate and Limits for Interval Age Data*: This explains how to estimate, plot, and interpret the sample MCF to obtain the information sought from interval data.

6. *Analysis of a Mix of Events*: This explains how to estimate, plot, and interpret the sample MCF for all events combined, for a single type of event, when some types of events are eliminated, and for selected groups of events.

7. *Comparison of Samples*: This shows how to compare two or more samples of recurrence data.

8. *Survey of Related Topics*: This includes various parametric models, reliability growth, renewal models, and models with covariates.

References.

Chapters 1–4 are basic and should be read first and in order. Other chapters may be read in any order.

Chapter overview. Chapter 1 contains the following sections, each of which describes a type of recurrent events data, corresponding data displays, and information sought from such data. For many applications, it is sufficient to read just sections 1.2 and 1.7:

1.2. *Exact ages and right censoring*.

1.3. *Exact ages, left censoring, and gaps*.

1.4. *Interval age data*.

1.5. *Continuous histories*.

1.6. *A mix of types of events*.

1.7. *Practical issues*. This section is essential background for applications.

Problems.

1.2 Exact Ages and Right Censoring

Purpose. This section presents the most common form of recurrence data. Data on discrete events consist of exact ages of recurrences and right censoring times. A discrete event is one that occurs at a single point in time. Three typical data sets illustrate such data: transmission repairs, bladder tumor recurrences, and costs of fan motor repairs on heat pumps. Each presentation describes the data, provides a display of the data, and describes information sought from the data. Analyses of the data appear in later chapters. These data sets are used to introduce basic concepts and various issues.

Transmission Application

Transmission data. A preproduction road test on a 1995 compact car was run at the Chrysler Proving Grounds. The test was accelerated by a factor of 5.5. That is, equivalent customer mileage is 5.5 times the test mileage. Table 1.1 lists repair data on automatic transmissions in a sample of 34 cars in a preproduction road test. The data consist of each Car ID, its test mileage at each transmission repair, and latest observed mileage (marked +). For example, car 027 was repaired at 48 and 1440 miles and was observed to 29,834 miles. Also, car 031 was observed to 21,762 miles without a repair. This and most other data sets in this book can be downloaded from http://www.siam.org/books/sa10/.

Table 1.1. *Automatic transmission repairs.*

Car	Mileage			Car	Mileage	
024	7068	26744+		115	17955+	
026	28	13809+		116	19507+	
027	48	1440	29834+	117	24177+	
029	530	25660+		118	22854+	
031	21762+			119	17844+	
032	14235+			120	22637+	
034	1388	21133+		121	375	19607+
035	21401+			122	19403+	
098	21876+			123	20997+	
107	5094	18228+		124	19175+	
108	21691+			125	20425+	
109	20890+			126	22149+	
110	22486+			129	21144+	
111	19321+			130	21237+	
112	21585+			131	14281+	
113	18676+			132	8250	21974+
114	23520+			133	19250	21888+

Age. Here transmission usage is measured in miles. In other applications, time is often the measure of usage. In general, *age* is the term used here to describe a unit's usage however it is measured.

Censoring. A unit's latest observed age is called its *censoring age* because the unit's event history beyond that age is censored (unknown) at the time of the data analysis. Usually, unit censoring ages differ. The different censoring ages complicate the data analysis and require the methods described in later chapters. A unit may have no events; then its data are just its censoring age. Other units may have one, two, three, or more events up to its censoring age. Also, censored data are called *truncated* data, and the censoring age is called the *truncation age* of the unit. This terminology is preferred by some authors, including Ascher and Feingold (1984) and Rigdon and Basu (2000).

Exact ages. Here the repair and censoring mileages of the cars all differ and are regarded as distinct points on the mileage scale. *Exact age data* consist of entirely distinct values on the age scale, no two values equal. This assumption simplifies the theory for analyzing such data. In practice, all age data are recorded to some number of significant figures, and some age values may be equal. Such age ties ordinarily have little effect on the results of analyses. However, Chapter 6 provides proper analyses for interval data, where ages are coarsely grouped into month, year, or some other relatively long time interval.

Display. Figure 1.1(a) displays the transmission data in a time-event plot. There the

1.2. Exact Ages and Right Censoring

data on each car are depicted on a straight line. The length of the line corresponds to the car's length of observation (censoring mileage). Each repair appears as an X at the corresponding mileage. Here the histories are ordered by car ID, which is order of production. This time-event display shows that cars produced earlier have more repairs than do later ones. This should be expected, as the production process was being debugged as the test cars were produced. Figure 1.1(b) displays the same histories ordered by censoring age, a display order often used here. This display does not reveal the information seen in Figure 1.1(a). These two displays show the value of making all kinds of data plots, as no one plot reveals everything. For example, the vertical scale in such plots can be any covariate of interest. Goldman (1992) and Lee, Hess, and Dubin (2000) survey varieties of such "event charts."

Information sought. Information sought from the data includes (1) the mean (cumulative) number of repairs per car by 24,000 test miles (equivalent to the design life of $5.5 \times 24,000 = 132,000$ customer miles) and (2) whether the population recurrence rate for repairs increases or decreases as the population ages. Analyses in Chapters 3, 4, and 7 provide this information. The automotive industry often uses the mean number of repairs per 100 cars by a specified mileage or age in days, denoted by $R/100$, which is like a percentage.

Bladder Tumor Application

Tumor data. Table 1.2 lists Byar's (1980) data on 87 recurrences of bladder cancer tumors in 48 patients receiving a placebo in a clinical study with two other treatments. Here the data on a patient consist of

- the patient's ID;
- the number of months the patient had been in the study at each recurrence; and
- the number of months the patient had been followed in the study, indicated by +.

For example, patient 10 had recurrences at 12 and 16 months and had been in the study 18 months, indicated by +. Also, patient 11 had been in the study 23 months with no recurrences. Here the data are recorded to the nearest month. Thus the age data are interval data but are treated as exact in subsequent analyses. A more proper analysis of such interval (grouped) data appears in Chapter 5. Data on all three treatments also appear in Andrews and Herzberg (1985). Other diseases with recurrent episodes include asthma, herpes, diabetes, epilepsy, and dental caries.

Display. Figure 1.2(a) displays these data and data on patients receiving Thiotepa treatment. In this display, the data on a patient is displayed on a time line, whose length equals the number of months the patient was followed, and each recurrence is denoted by an X (Thiotepa) or ■ (placebo) at the corresponding number of months. For example, the line for placebo patient 27 is 31 months long and has ■s at 12, 15, and 24 months. Also, the line for patient 11 is 23 months long and has no ■s. This display was made using the JMP software of the SAS Institute (2000).

History function. Figure 1.2(b) shows the observed *cumulative history function* for the number of recurrences for patient 27. It is an alternative depiction of recurrence data. It shows the cumulative number of recurrences (vertical axis) as a function of the number of months in the study (horizontal axis). From 0 to 12 months, the cumulative number of recurrences is flat at zero. At 12 months, the cumulative number of recurrences jumps to one and continues flat to 15 months. At 15 months, the cumulative number jumps to two recurrences and continues flat to 24 months. At 24 months the cumulative number jumps to

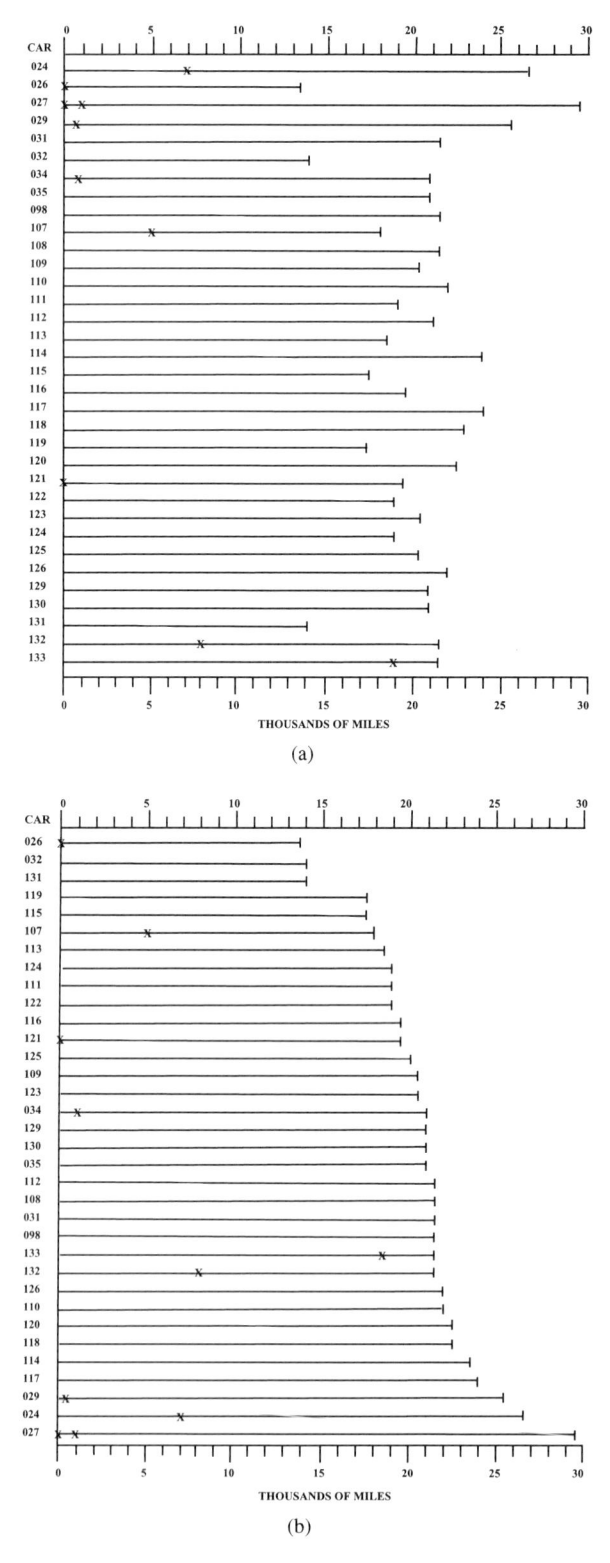

Figure 1.1. (a) *Display of automatic transmission data in production order.* (b) *Display of transmission data in order of censoring age.*

1.2. Exact Ages and Right Censoring 7

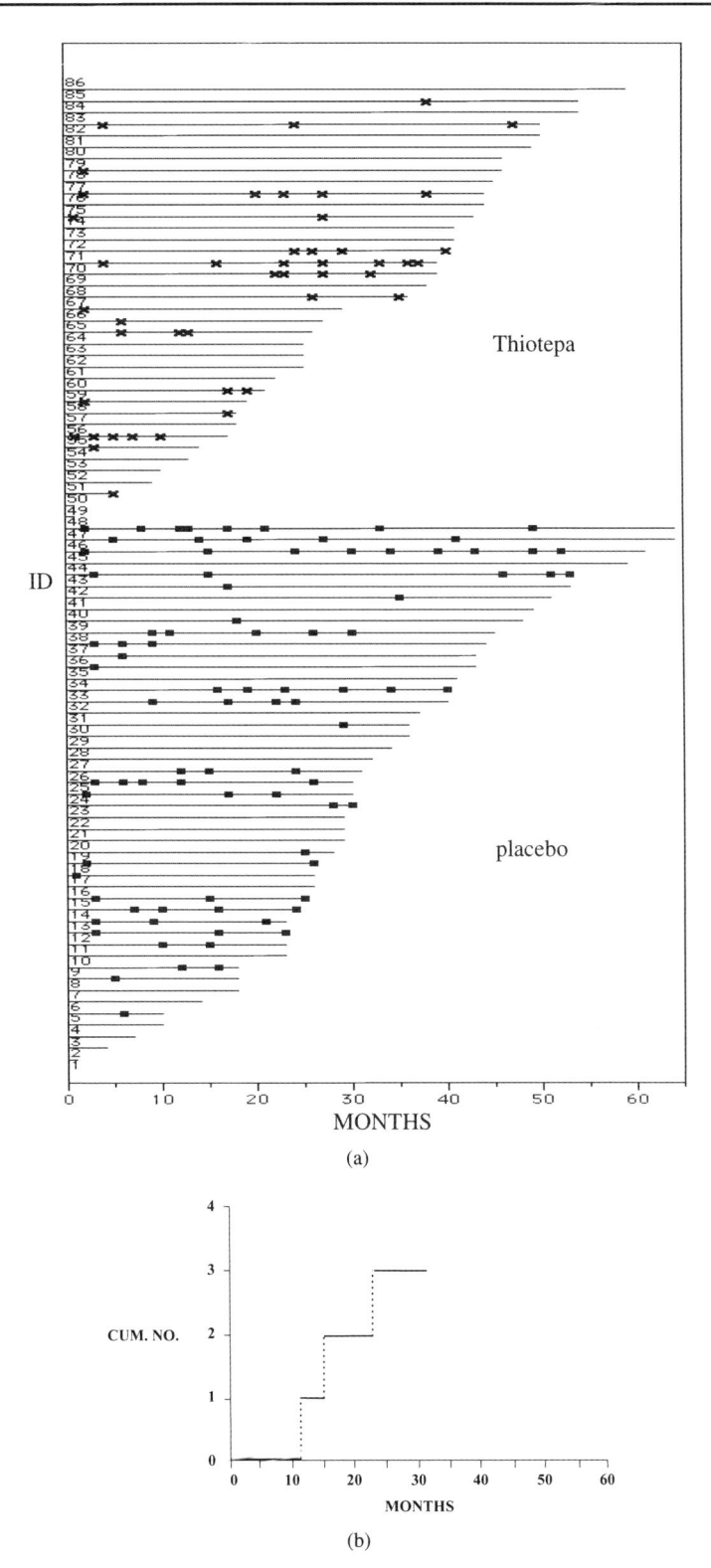

Figure 1.2. (a) *JMP display for placebo (■) and Thiotepa (X) treatments.* (b) *Cumulative history function for the number of recurrences.*

Table 1.2. *Tumor recurrence data for placebo treatment.*

ID	Months	ID	Months
1	0+	25	2 17 22 30+
2	1+	26	3 6 8 12 26 30+
3	4+	27	12 15 24 31+
4	7+	28	32+
5	10+	29	34+
6	6 10+	30	36+
7	14+	31	29 36+
8	18+	32	37+
9	5 18+	33	9 17 22 24 40+
10	12 16 18+	34	16 19 23 29 34 40 40+
11	23+	35	41+
12	10 15 23+	36	3 43+
13	3 16 23 23+	37	6 43+
14	3 9 21 23+	38	3 6 9 44+
15	7 10 16 24 24+	39	9 11 20 26 30 45+
16	3 15 25 25+	40	18 48+
17	26+	41	49+
18	1 26+	42	35 51+
19	2 26 26+	43	17 53+
20	25 28+	44	3 15 46 51 53 53+
21	29+	45	59+
22	29+	46	2 15 24 30 34 39 43 49 52 61+
23	29+	47	5 14 19 27 41 64+
24	28 30 30+	48	2 8 12 13 17 21 33 49 64+

three recurrences and continues flat to 31 months, the length of observation. This staircase function is typical of counts of recurrences. The dotted vertical lines are mathematically not part of the history function; they merely guide the eye and connect the steps.

Information sought. The study was conducted to compare how the three treatments affect tumor recurrence. In Chapter 7, just two treatments are compared. Also, the investigators sought to understand how the rate of tumor recurrence behaves over time, that is, the course of the disease. This information is an aid to counseling patients and scheduling examinations. Analyses in Chapters 3, 4, and 7 provide this information.

Fan Motor Cost Application

Fan motor costs. Table 1.3 shows repair cost data on 119 residential heat pumps on a service contract that experienced 26 fan motor repairs. The data consist of

- the heat pump age in days at each repair and corresponding repair cost, denoted by the pair (age, cost), and

- the latest (censoring) age of each heat pump, denoted by (age +).

The censoring ages of 73 heat pumps are omitted; all are slightly over 3652 days (about 10 years). The repairs and censoring ages are not identified with particular heat pumps. This loss of information has consequences described in Chapter 4. In particular, in this circumstance one cannot calculate proper confidence limits for estimates. When one knows the unit ID for

1.2. Exact Ages and Right Censoring

Table 1.3. *Fan motor cost data.*

Days	Cost ($)	Days	Cost ($)	Days	Cost ($)	Days	Cost ($)
141	44.20	1170+		2363	225.40	3296	145.00
252+		1200+		2419+		3307	208.00
288+		1232+		2440	197.80	3368	248.70
358+		1269	130.20	2562+		3391	256.90
365+		1355+		2593	184.00	3406+	
376+		1381	150.40	2615+		3440	305.50
376+		1471	113.40	2674	167.20	3489	202.80
381+		1567	151.90	2710	149.00	3490+	
444+		1642	191.20	2794+		3621+	
651+		1646	36.00	2815+		3631+	
699+		1762+		2838	42.00	3631+	
820+		1869+		2946	255.70	3631+	
831+		1881+		2951	243.30	3631+	
843	110.20	1908	158.50	2986+		3631+	
880+		1920+		3017+		3635	242.00
966+		2110+		3087+		3639+	
973+		2261	243.80	3216+		3648+	
1057+		2273	189.50	3291+		3652+	

(73 other fan motors have no repairs and have censoring ages just above 3652+ days.)

each event, the data on a unit has the form ID, (age 1, cost 1), (age 2, cost 2), ..., (age K, cost K), (censoring age +).

Display. Lacking heat pump IDs, these data cannot be displayed as before with a line for each heat pump. At best one could display the 26 repair ages and 119 censoring ages on a single time scale. The cost could be displayed numerically alongside each repair age.

Information sought. The service contract provider wanted to know how such costs accumulate over time. The provider used this information to price service contracts, to forecast costs, and to decide which brands of heat pump to exclude from service contracts. Analyses in Chapters 3 provide this information.

History function for cost. Usually all the repairs and costs and the censoring age associated with each unit are known. Then a unit's cumulative history function for cost has the appearance shown in Figure 1.3. Here the rises in the staircase function for *costs* are not all equal; instead each step height equals the corresponding cost. In contrast, in Figure 1.2(b) for

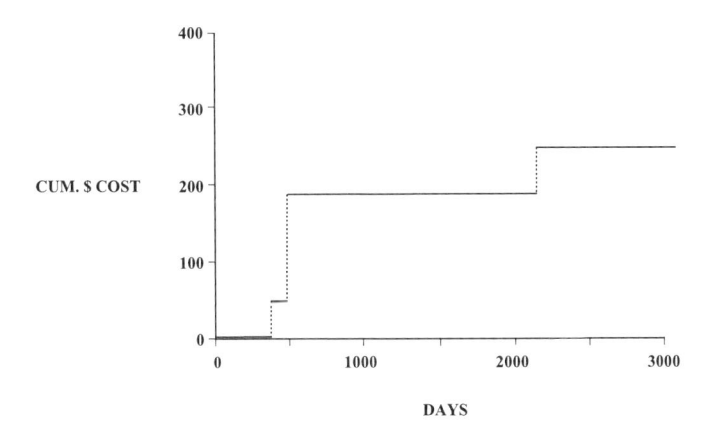

Figure 1.3. *Cumulative history function for cost.*

the cumulative *number* of recurrences, all step heights are usually one (recurrence). Patient treatment costs for most diseases can be viewed with such cumulative cost histories. Cost functions may have negative steps. For example, a bank account balance increases with each deposit and posting of interest and decreases with each withdrawal. Similarly, the debt of a borrower increases with each added loan and interest posting and decreases with each payment. Morin (2002) reports that the MCF for cost has been used to study the purchasing behavior of Internet shopping customers.

Value data. Recurrence data can have some numerical value other than cost associated with each event, for example, the downtime to repair a failed system. Another example is the number or size of bladder tumors at each recurrence. Such a numerical value is a measure of the size, importance, severity, performance, etc., of an event. In general, "cost" will be used to mean any such quantitative measure. In the literature, such data are called "marked point processes." Also, data can have more than one numerical variable associated with each event, for example, repair cost and downtime. Such performance data are then multivariate. Data with a mix of types of events (Chapter 6) involve a count for each type and thus are multivariate.

1.3 Exact Ages, Left Censoring, and Gaps

Purpose. Data with exact event and censoring ages can be censored on the left and can have gaps as described below, a less common situation. Such data are illustrated with compressor replacements on heat pumps. This section describes these data, provides a display of the data, and describes information sought from the data. The data are analyzed in Chapter 3, section 3.4.

Compressor data. Ross (1989) gives data on 28 replacements of heat pump compressors under service contract in commercial buildings. Table 1.4 presents data from five buildings (B, D, E, H, K). Buildings H and K started on service contract when the heat pumps were installed (age 0). The 164 compressors in Building B started on service contract at 2.59 years after installation; their preceding replacement records were unavailable, that is, *left censored*, also called *left truncated*. These 164 compressors enter the sample at this age and are denoted in Table 1.4 by 2.59 (+164). Similarly, Building D has 356 compressors left censored at 4.45 years, and Building E has 458 compressors left censored at 1.00 years. For each of the 28 compressor replacements, the corresponding heat pump age in years appears in the column for its building. For example, for Building B, the first replacement occurred at 3.30 years and the last at 9.27 years. The (right) censoring age of heat pumps in Building B is 9.33 years. This appears in Table 1.4 as 9.33 (−164); the minus indicates that 164 heat pumps are censored at that age. Censoring ages for heat pumps of other buildings similarly appear in the table. All censoring ages are boldface for clarity.

Display. Figure 1.4 displays the compressor data versus age in years. There each building history appears as a straight line segment that extends from its (left censoring) age when observation started to its latest (right censoring) age. Periods of missing data (gaps) are indicated with dotted lines. The display has an X at each compressor replacement age. The display also shows the number of heat pumps at the start and end of a building's history. For example, Building B has +164 compressors at the start and −164 at the end. The + and − notations are explained in Chapter 3, where these data are analyzed.

Information sought. The investigator sought to estimate the compressor replacement rate and to understand its nature, i.e., whether it increases or decreases as the heat pumps age. Also, he sought the median life of such compressors. Chapter 3 provides analyses of these data.

1.4. Interval Age Data

Table 1.4. *Compressor repair age data in years.*

Building B	Building D	Building E	Building H	Building K
2.59 (+164)	**4.45** (+356)	**1.00** (+458)	**0.00** (+149)	**0.00** (+195)
3.30	4.47	2.58	0.17	2.17
4.62	4.47	4.65	0.17	3.65
4.62	5.56	4.79	1.34	4.14
5.75	5.57	5.85	**5.09** (−149)	**4.14** (−195)
5.75	5.80	6.73		
7.42	6.13	**7.33** (−458)		
7.42	7.02			
8.77	**7.05** (−356)			
9.27				
9.27				
9.33 (−164)				
10 replaced	7 replaced	5 replaced	3 replaced	3 replaced

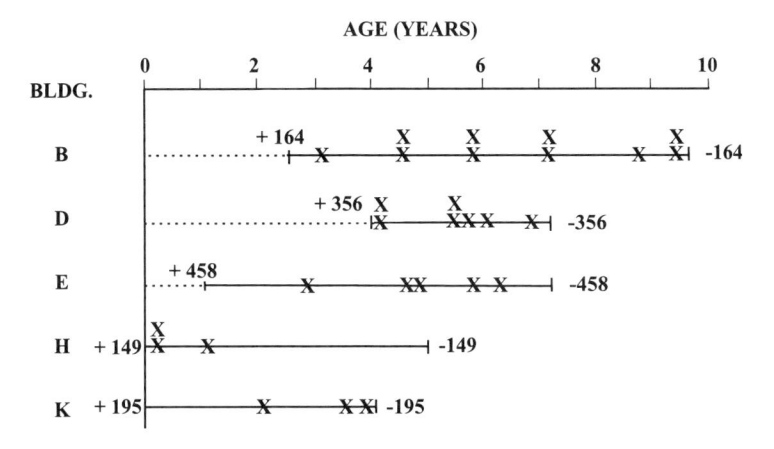

Figure 1.4. *Display of compressor data.*

Gap data. Thus far, the applications have concerned data censored on the right or on the left. There are other applications where each unit is observed over some time intervals and not observed over others. For example, drug addicts in a recovery program were tested weekly for drugs in their blood. Of course, some addicts missed their tests from time to time, producing gaps in their records. The age intervals where units are not observed are called "gaps." Right censoring is a gap to the right of the last censoring age. Left censoring is a gap from age 0 to the first observation age.

Extensions. The compressor example above concerns the *number* of repairs. The methods in Chapter 3 for exact age data with left censoring and gaps apply to *cost* and *value* data on recurrences as well.

1.4 Interval Age Data

Purpose. This section describes interval data on discrete events. Here one does not know the exact event ages and censoring age for a unit. Instead, the age scale has been partitioned into intervals, and one knows only the number of events and the number of censoring ages in each interval. Such grouping of data into intervals is usually done to compress large data sets. For example, the defrost control data set in Problem 5.2 has 22,914 controls. Grouping loses some information, but usually not enough to matter. Interval data are illustrated here with births of children to statisticians. This section describes the data, shows a typical data display, and describes information sought from the data. Analyses for such data appear in Chapter 5.

Childbirth data. Childbirth data were collected from statisticians attending the author's short course on recurrence data analysis for the Southern California chapter of the American Statistical Association (ASA). The data were used to illustrate how to analyze grouped recurrence data. On a questionnaire each attendee provided his/her sex, age at the birth of each child, and current age. For example, the author's data then were male, 24, 26, 28, 61+, which are his age at the birth of each of his three children and his current age, labeled +. Data on an attendee with no births is just his/her sex and current age, say, female, 31+. The number of children born to the attendees ranged from 0 to 8.

Display. Figure 1.5 is a display of such grouped recurrence data for just six individuals. Each is depicted with a dashed horizontal line. Each dash shows the length of the corresponding interval, all two years in this figure. The length of the line shows how long the individual was observed. The censoring age is indicated with a C in the corresponding interval. The birth of each child to a parent is denoted with an X in the corresponding age interval. Multiple births are denoted with 2 (twins), 3 (triplets), etc.

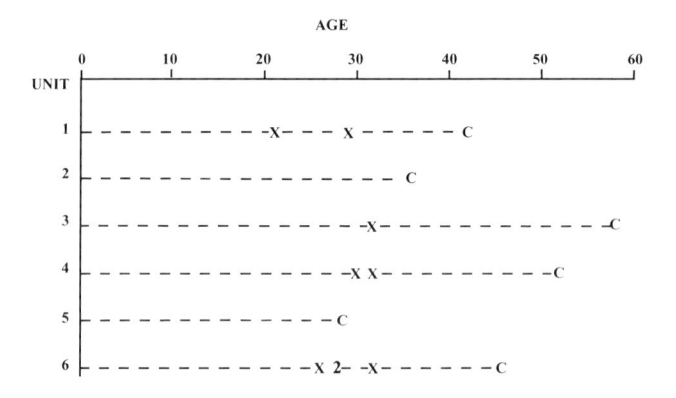

Figure 1.5. *Display of typical histories, interval data.*

Grouping. The data are summarized separately for men and women in Table 1.5. The 104 attendees consisted of 45 women with 45 offspring, and 59 men with 63 offspring. To condense the data, the ages at childbirth were grouped into convenient intervals of two years (18–20, 20–22, etc.), five years (40–45, 45–50, etc.), and 39 years (60–99). Intervals need not have equal length, but all individuals must be grouped into a common set of intervals. For each age interval (columns 1 and 2), the table shows the number of children (columns 3 and 5) born to attendees in that age interval and the number of attendees (columns 4 and 6) with current ages in that interval. Thall and Lachin (1988) deal with more complex data where different individuals can have different intervals.

Information sought. The data analyses in Chapter 5 provide information on the following questions:

- How do the childbirth rates of male and female statisticians compare as a function

1.5. Continuous Histories   13

Table 1.5. *Childbirth data.*

Interval i	Age Endpoints $t_{i-1} - t_i$	Men Number of births R_i	Men Number censored C_i	Women Number of births R_i	Women Number censored C_i
0	0–20				
1	20–22	3	1		
2	22–24	3		1	1
3	24–26	3	1	6	2
4	26–28	5	1	8	7
5	28–30	6	3	6	3
6	30–32	11	3	9	3
7	32–34	6	5	7	3
8	34–36	9	2	3	2
9	36–38	6	3	5	1
10	38–40	2	3		1
11	40–45	7	7		9
12	45–50	2	14		3
13	50–55		5		4
14	55–60		8		4
15	60–99		3		2
	Total:	63	59	45	45

of age? Common experience suggests that women tend to start having children at a younger age than men do. Also, experience suggests that men continue having children into older ages since women are limited by menopause.

- How many children on the average do statisticians ultimately have? To sustain a population, this average needs to be about 2.1. Two children are needed to replace the statistician and mate, and the excess 0.1 accounts for children who die before childbearing age.

- Do men and women ultimately have the same average number of children? Of course, common knowledge provides an answer to this.

Give thought to what you expect the data to show.

Extensions. The childbirth application concerns the *number* of recurrences. Other applications with interval age data can involve cost or "size" data. Then one tabulates the *total cost* of all recurrences in an interval in place of the *total number* appearing in Table 1.5. Also, data may be left censored and have gaps.

1.5 Continuous Histories

Purpose. In some applications, a cumulative performance variable for a unit varies continuously over time. Such applications are described here. The methods for analyzing data on discrete events extend to such continuous histories, as described in later chapters.

Continuous histories. In the previous applications, the events were all *discrete*. That is, each event happened at a single point in time. For example, transmission failures and childbirths are regarded as happening at a certain point in time. In some applications, the cumulative value $Y(t)$ of a unit is a continuous function of time t, for example, the cumulative energy output of a power plant, degree-days, rainfall, and sunshine. Figure 1.6 shows such a cumulative history or curve. The derivative (slope) $dY(t)/dt$ at time t is the plant's power output at that instant. The figure shows that the plant's power level was constant at different levels over different time intervals. Over one time interval, the slope is 0, indicating that the plant produced no power then. Other examples include the total energy output of a locomotive versus time or miles, the cumulative cost of heating and cooling a facility over time, the cumulative degree-days used as a basis for such heating costs, or the cumulative milk output of a cow over time.

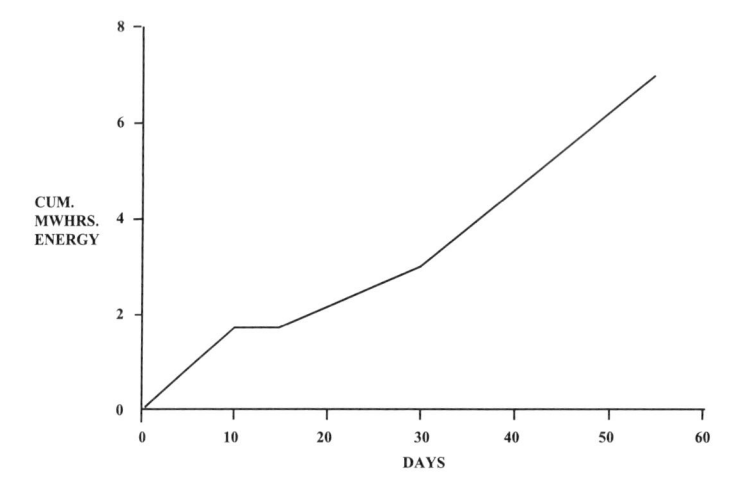

Figure 1.6. *Cumulative energy output of a power plant.*

Availability. Power plants, locomotives, production lines and equipment, and many other systems are evaluated with respect to availability. Roughly speaking, a system's *availability* is its actual running time (uptime) divided by its scheduled running time. Equivalently, for a "fleet" of systems, fleet availability is the actual number of running systems divided by the scheduled number. For availability purposes, a system history function is the cumulative uptime (say, in hours) as a function of scheduled running time (in hours). Figure 1.7 depicts such a history function. When the system is up, the slope of its function is 1, as it accumulates one hour of uptime per hour of scheduled running time. When the system is down, the slope is 0.

Defect growth. A common application having continuous histories concerns the incubation, initiation, and growth of a defect in a unit. For example, a mechanical component or fatigue specimen initially has no crack; after a time a crack "initiates" (is detected with the method of observation) and then grows over time, possibly to component failure. Figure 1.8 depicts such a history as a curve.

Information sought. A sample of such cumulative history functions is used to model, understand, and predict population behavior. For example, the information may be used to schedule inspections or preventive replacement of jet engine components. Analyses for such continuous sample histories appear in Chapters 3 and 4.

Extensions. Such "continuous" history functions may also have jumps up or down. Also, they can be censored on the right and left and have gaps. As depicted here, such history functions are *continuously observed* to yield such curves for every point age in the period observed. In other applications, the history functions may be *discretely observed* at a number of discrete points in time. Then one knows only the increments in a function

1.6. A Mix of Types of Events

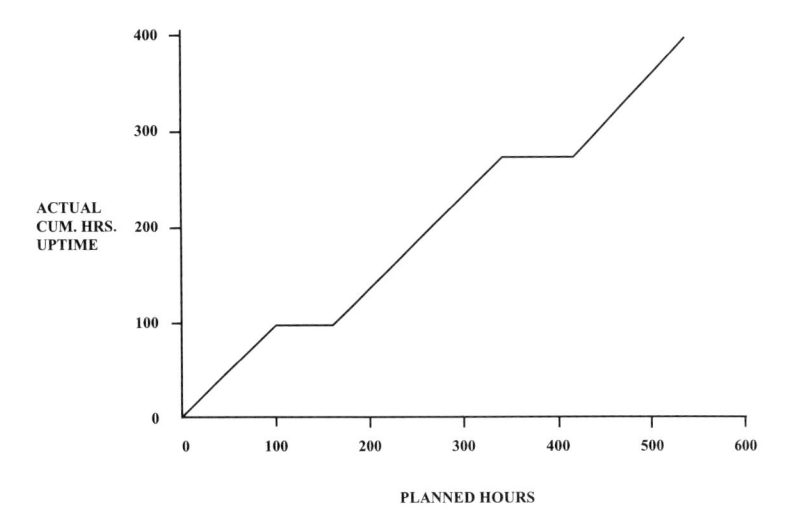

Figure 1.7. *Actual cumulative running time versus planned time.*

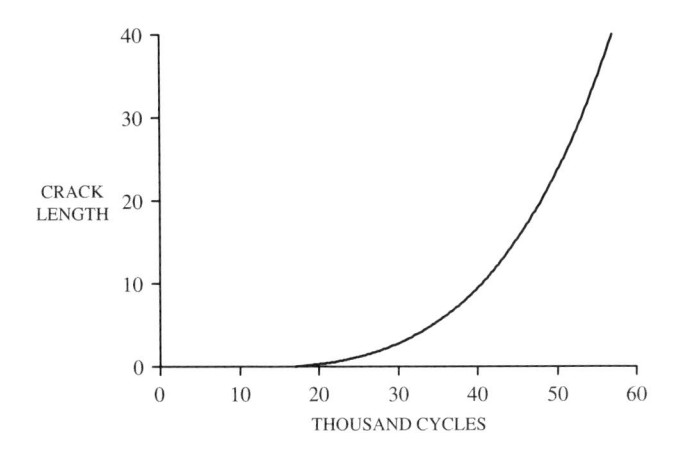

Figure 1.8. *Crack initiation and growth.*

between successive observation ages, and the data are interval data, also called inspection data. For example, jet engines are opened and inspected periodically to observe the size of cracks in noncritical components. In some applications, there is only one inspection per unit. For example, when a critical component fails in a jet engine, the entire fleet is inspected for that component. Then crack size is observed once on each engine and each has a different inspection age.

1.6 A Mix of Types of Events

Purpose. All previous applications are concerned with just one type of event or all types are pooled. This section describes data with a mix of types of events. For example, data on systems may contain any number of failure modes. Also, childbirths could be classified as male or female. For such data, one may wish to estimate the MCF for any of the following:

- *All types of events combined*: This yields the total recurrence rate or cost.

- *A single type of event*: This yields the MCF of that event as if it were the only one. An example is a single failure mode of a product.

- *A group of types of events*: This yields an MCF for that group. For example, one may be interested in all the failure modes of a particular component or subsystem. This shows how much that component or subsystem contributes to the system MCF. This information is used to allocate improvement efforts to failure modes, components, or subsystems.

- *The events left after some are removed*: In working on a product, engineers often want to know how much the number or costs of repairs would be reduced if certain failure modes were eliminated through improved design. This indicates whether such improvements are worthwhile.

Traction motor data. Table 1.6 displays service data with 164 traction motor failures (21 failure modes) on 372 subway cars, each with four motors. These interval age data are given in one-month intervals from one to 33 months. Each column of the table shows the data on a failure mode (denoted with a letter, A, B, etc.). Each row shows a particular month in service. The number of failures by a particular mode in a particular month appears in the column for the mode and the row for that month. For example, there were four failures of mode O in month 16. The "Censored" column shows the number of cars censored in each month, that is, the number of cars at each monthly age. Also, the total number of events in each row (month) appears in the far right column, and the total number of events in each column appears in the bottom row. For example, the total number of censoring ages is 372, which equals the number of cars. Also, for example, the total number of mode A failures is 26, and the total number of failures of all types in month 10 is seven.

Information sought. The manufacturer sought information for two main purposes: (1) to evaluate the probability that a new fleet to be sold to a customer would pass a reliability demonstration test in service and (2) to identify how best to improve the current design. In particular, the manufacturer sought the MCF estimate for the following:

- *All failure modes combined*: This yields information on the probability of passing the demonstration test.

- *Each failure mode separately*: This shows how much each mode contributes to the total MCF and identifies which failure modes are most in need of improvement.

- *Groups of failure modes*: The failure modes were assigned into three groups: modes in (due to) the design, modes outside the design, and an unassigned mode K. The MCF estimates indicate where redesign effort is needed.

- *Modes eliminated*: It was thought that modes A, B, and C could be eliminated through redesign. Would the reduction in the MCF be worthwhile?

Analyses that yield this information appear in Chapter 6.

Display. Table 1.6 is a rough but simple display of interval data with a mix of types of events. It suffices as a display for many applications. Figure 1.9 is a time-event display of such sample data.

Event types. In practice, one must decide how to classify observed events into types. This partitioning of the events can be done in any unambiguous way that is useful in the

1.6. A Mix of Types of Events

Table 1.6. *Traction motor data.*

Month	Censored	D	N	F	O	H	K	B	C	G	L	A	Q	R	E	I	J	M	S	X	P	T	All
1		2										1											3
2			1																				2
3					1																		2
4		1													1								5
5																		2					2
6		1																					3
7																							6
8	13																						6
9	28																						4
10	28											1										1	7
11	29				2							5						2					6
12	27			1	1		2					1	1										5
13	25			1	2		2						1		2								12
14	24	1	1				1																4
15	1		1	2							1	1					1						5
16	11		2	1	4		4						1						1				13
17	7			1	1		1				3	3											9
18	12	1		1	2						2	3	1							1			8
19	14	1			1		2				4	1						1					11
20	12	1		1		2															1		6
21	14	1		1			1					3				1							3
22	11	1	1				1	1					1										6
23	13	1			1							1	2							1			3
24	9	1	1		2		4					1	1										13
25	8				3					2													9
26	12	1				1			1														3
27	13			1	2		1							1									5
28	11	1		1																			2
29	9												1										1
30	13																						
31	13																						
32	12																						
33	3																						
Sum:	372	12	7	8	23	3	24	3	3	2	19	26	9	2	3	1	2	11	2	2	1	1	164

```
                        AGE
          0   10   20   30   40   50   60
          |....|....|....|....|....|....|
    UNIT
     1    |----------A--C------|
     2    |-----------------|
     3    |-C--------------B-----------|
     4    |---------------CB--------|
     5    |--------------|
     6    |-----------C-A-B-----|
```

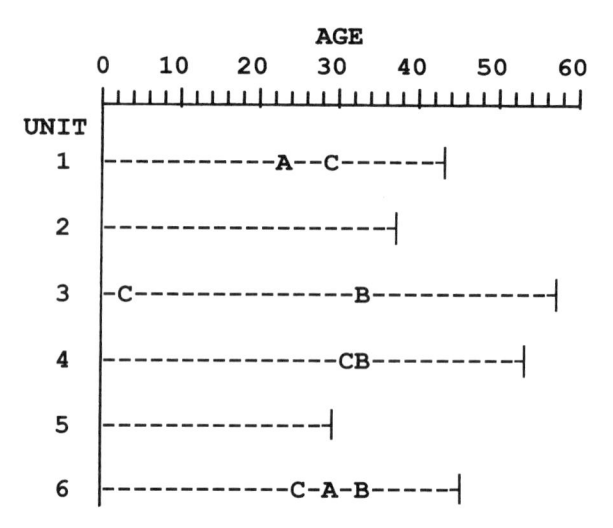

Figure 1.9. *Display of sample histories with a mix of events.*

application. For example, for the traction motors there are 21 failure modes. Also, one could group them into three events: in the design, outside the design, and unknown. Thus one can use more than one classification of types of events. Failure modes may be physically dependent, that is, resulting from a common underlying cause. Then it may be useful to group them together and treat them as a single mode.

Extensions. Such data extend to the following situations:

- The traction motor application has interval age data. One can also have exact age data with a number of types of events.

- One can analyze cost (or value) data, instead of the number of events.

- The age data may be left censored or have gaps.

- Value histories may be continuous functions of age.

Turbine data. In another application, data on repairs on a type of naval turbine in a fleet were collected. The primary purpose of the data analysis was to determine which failure modes and repairs most affected turbine availability, maintenance costs, and downtime. This information would be used to guide the design of the next version of such turbines. Chapter 2 presents the pooled turbine data (all failure modes).

1.7 Practical Issues

Purpose. This section reviews a number of practical issues that must be resolved by practitioners in the field of application. In contrast, statistical issues are covered in Chapters 3 and 4. While many of these issues are described here in engineering terms, the same or similar issues arise in most other fields. Lawless (1998), Robinson and McDonald (1991), and Suzuki, Karim, and Wang (2001) discuss some of these and other issues for warranty claim data.

1.7. Practical Issues

Population. Data are collected to provide information on some target population. In practice, the target population should be precisely specified and the data should be representative of that population. For example, for a manufactured product, the target population is typically all units made with a particular design. For the study of bladder tumors, the target population is likely all people with the disease. Ideally, a sample should be taken randomly from the target population. In practice, many samples are not truly random. The sample, a statistical issue, is discussed at length in Chapter 3. The childbirth data for statisticians do not have a clear target population (what is a statistician?), and the nonrandom sample consists of people who chose to participate in a course.

Events. It is essential to have a clear definition of what is and is not an event or recurrence. For example, is a miscarriage a birth? If an appliance is found working during a service call, is it a failure?

Repair/failure. In reliability engineering work, there are many possible definitions of repair or failure, the event/recurrence of interest. Each definition is appropriate for some purpose, and there is no one correct universal definition. In practice, data may be repeatedly analyzed with different definitions, each yielding some insight for engineering, management, accounting, marketing, and others. Examples follow:

- Many failures result in a product that does not function at all, a clear and obvious failure.

- The performance of other products degrades to the point that the owner complains. Similarly, in an engineering test, a defined failure occurs when product performance degrades below some reasonable but arbitrary level.

- A percentage of replaced components, when returned and examined by engineering, work properly. In some cases, these are intermittent failures. For engineering purposes, usually only true physical failures are of interest. For management purposes, any *replacement* has incurred a cost and possible customer dissatisfaction and is thus of concern.

- In some applications, not all product failures are reported to the manufacturer. Should the unreported ones somehow be taken into account? Does it make a difference whether such failures are during warranty or beyond?

- There is a range of complaints, repairs, and service calls, any of which could be included or excluded as an event of interest. These include the following:

    - The serviceman finds that the owner has not plugged in the product.
    - The product owner has misused the product and must be instructed on proper use.
    - The problem is not solved until the second or third service call.
    - The owner complains that the product does not perform as expected, is noisy, etc.
    - The product needs an adjustment.
    - A component is repaired, for example, resoldered.
    - One or more failed components are replaced.
    - Preventive replacement of some components is performed.
    - Scheduled maintenance and related costs can be included.

Some reliability literature is concerned with the state of a system after repair, for example, "good as new" and "bad as old" (which means as good as others of the same age). Section 8.4 deals with renewal models where repaired systems are restored to new condition. Kvam, Singh, and Whitaker (2002) and their references present repair models and data analyses where units are not restored to new condition. The models and statistical methods in this book are valid for any definition and type of repair.

Age. "Age" (or "time") is used here to mean any appropriate measure of system usage. For a heat pump, age is the number of days since installation. Other possible measures of heat pump usage are its running hours and its accumulated energy output, but they usually are not available. For a car, mileage is the usual measure of usage, but the automotive industry uses days in service for many purposes. Some products go through a usage cycle, for example, an X-ray machine or blood sample analyzer. Their usage is the number of cycles. In the aviation industry, a cycle of takeoff and landing is a key measure of usage for jet engines, rather than flying hours. For locomotives, the industry has used days in service, mileage, and horsepower-hours (energy output) as measures of usage.

Downtime. Downtime during system repair may be included in or excluded from the usage, according to whatever is appropriate for the application. With days in service as the usage, downtime is often included; this is the practice for the residential heat pumps and cars. For blood analyzers, cycles are counted and do not accumulate during repairs. For marine turbines, unscheduled downtimes are included and used to calculate availability.

Time 0. For most applications, there is a reasonable time 0, when the history of a unit begins. For example, for the birth data, time 0 is when a sample statistician was born. Also, for most products, time 0 is when the unit is purchased, delivered, or installed. To get this date, many manufacturers of consumer appliances request that the consumer send in a card with the purchase date (often getting a poor response rate). In contrast, for other applications, there is no obvious time 0. For example, for patients in a disease study, possible choices are date of first symptoms, date of diagnosis, date of reaching a specific stage of the disease, and date of entry into the study (the choice for bladder tumors); Andersen et al. (1993, Chapter X) discuss this issue. For products whose date of installation is not known, common choices are date of manufacture and date of sale. The choice of time 0 is often dictated by the information that is available.

Which age? In some applications, there is a choice of the age at which an event is said to occur. For example, one may record the age at which a product failed, when its failure report arrived, or when its repair was completed. Similarly, onset of a disease may be when the patient first observed symptoms, when the patient was diagnosed, or when the patient entered the study. In tracking medical costs over time, the choice could be the date of treatment, the date of billing, or the date of payment. Often the choice is dictated by the information that is available.

Multiple usages. For some products, failure and repair may be related to more than one measure of usage. Then both measures of usage may be used to model repair data. For example, the recurrence rate of repairs on gas turbines depends on both the number of startups and the number of running hours. Also, the recurrence rate of repairs on certain blood analyzers depends both on how many samples they analyze (cycles per day) and the number of hours they are on per day, as some components fail from wear and others from degradation at high temperature. Nelson (1995b) presents a model and data analyses for this situation. Similarly, the automotive industry uses mileage and days in service for some analyses. Statistical methods for such automotive data appear in Lawless, Hu, and Cao (1995) and Suzuki (1993).

Terminations. The statistical models and methods in this book require that each unit's unknown future continue in principle beyond its current censoring age to any age of interest. In contrast, in some applications, a unit's history terminates and cannot continue beyond its termination age. For example, a patient may die while in a medical study. Also, a product may be scrapped or retired. In such cases, we know the entire future of such units—they have no further recurrences. Such cases have informative censoring and require special treatment.

Costs. Practitioners must decide what is to be included in dollar costs. For example, for repairable products, one can include any of the following: purchase price, financing charges, scheduled maintenance, repairs, service, energy and other consumables, depreciation, and (negative) trade-in value. Different combinations of such items can be analyzed, each yielding different insight. Also, one can use actual dollars or convert them to constant dollars or to future value. In Morin's (2002) study of the purchasing behavior of Internet shoppers, the cost or profit of a purchase could be used.

Other values. As noted above, one can choose any useful measure of the value, size, importance, severity, etc., of an event.

Errors. Data collection and analysis are subject to errors from many sources. These must be taken into account and minimized and include the following:

- The sample is not representative of the population of interest (biased sample). This happens when the sample is not truly random, when some sample units are not from the population, or when some of the population cannot get into the sample. For example, prototype units are tested in the lab to predict the performance of production units in service.

- Some events are unreported and others reported incorrectly (reporting error).

- Some ages may be incorrect for various reasons (reporting error).

- Cost and other values are measured incorrectly (measurement error) or misrecorded (recording and transcription error).

- The model fitted to the data is inadequate (model error).

Problems

1.1. Your data. For a data set of your choosing, do the following:

(a) Make a display (time-event plot) of your data. Arrange the units in any meaningful order in the plot. Describe what the plot shows.

(b) Plot the cumulative history functions for some or all of your sample units. Describe what you see.

1.2. Bibliography. Assemble a bibliography of applications in your area.

Chapter 2
Population Model, MCF, and Basic Concepts

2.1 Introduction

Purpose. This chapter presents the basic nonparametric model for a population. This model is used throughout to analyze recurrent events data. A key feature of this population model is its mean cumulative function (MCF), which yields most information sought from such data. Basic parametric models are surveyed in Chapter 8.

Chapter overview. This chapter contains the following sections:

2.2. *Cumulative history functions*: This reviews typical cumulative history functions for the population model.

2.3. *Population model and its MCF*: The model is the uncensored population of all cumulative history functions. Its MCF contains most information sought from such data.

2.4. *Information sought*: This reviews types of information that can be obtained from the MCF.

Problems.

This is all basic material and is essential background for all later chapters.

2.2 Cumulative History Functions

Purpose. This section presents *population* cumulative history functions. These include functions for counts, costs, and a continuously varying performance value.

Censoring. Chapter 1 presented *sample* cumulative history functions, which are censored. Censoring is a property of the data collection and is not a property of the population. Thus for the population, a cumulative history function for any unit is regarded as uncensored over any potential age range of interest.

Count histories. Figure 2.1 shows a typical censored sample cumulative history function for the *number* of recurrences of discrete events (i.e., each occurs at a single point in time). That function, which potentially could be observed to any age of interest, is the "value" of that population unit. Thus the population "value" for each population unit is not a single number but is a *curve*—in particular, a staircase function as in Figure 2.1. In some parametric models, like the Poisson process, the increments are all 1, as simultaneous events

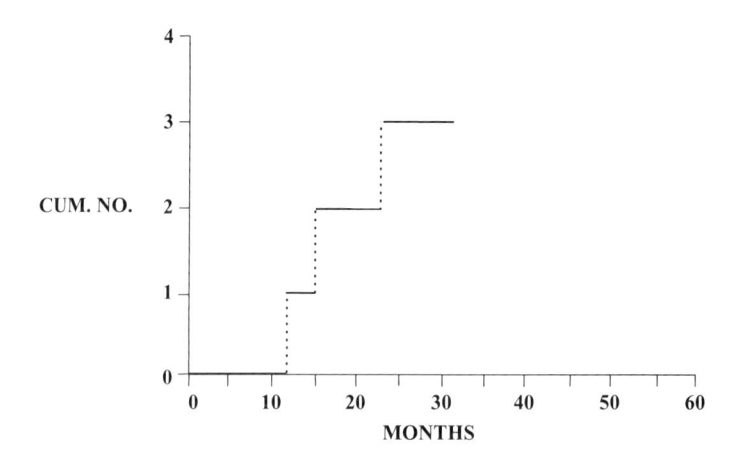

Figure 2.1. *Cumulative history function for the* number *of recurrences.*

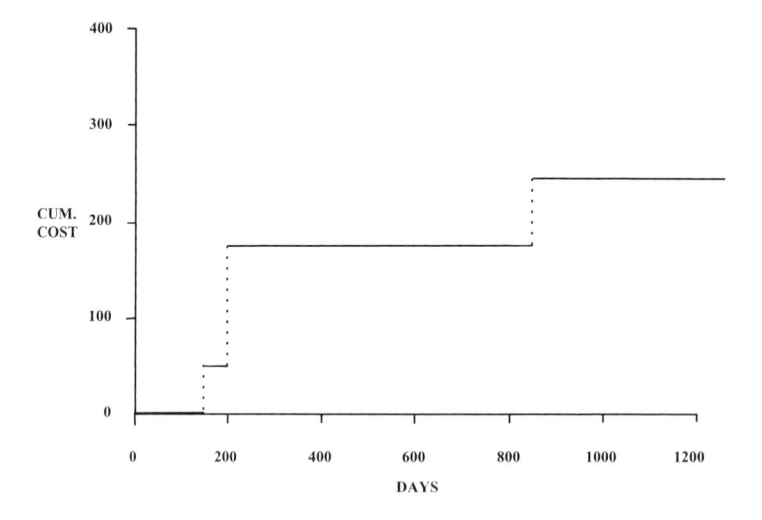

Figure 2.2. *Cumulative history function for cost.*

are impossible in that model. More generally, the increments can be any integer 1, 2, 3, ... For example, the births data could contain twins, triplets, etc. The general model here also allows for negative increments. For example, the number of people or production units in a waiting line can increase and decrease. Hereafter, the words "cumulative," and less often "history," may be omitted when referring to such functions, but they are implied.

Cost/value histories. Figure 2.2 shows a typical censored sample cumulative history function for the *cost* of recurrences. That function, which could potentially be observed to any age of interest, is the "value" of that population unit. Thus the "value" for each population unit is not a single number but is a *curve*, a staircase function with unequal step rises, as in Figure 2.2. As before, any measure of the "value" of an event may be used, for example, downtime of a unit during repair. Also, such history functions may have negative increments, for example, withdrawals from a bank account or customer returns of purchases.

Continuous histories. Figure 2.3 shows continuous cumulative history functions. Such a continuous function, which could be observed to any age of interest, is the "value"

2.3. Population Model and Its MCF

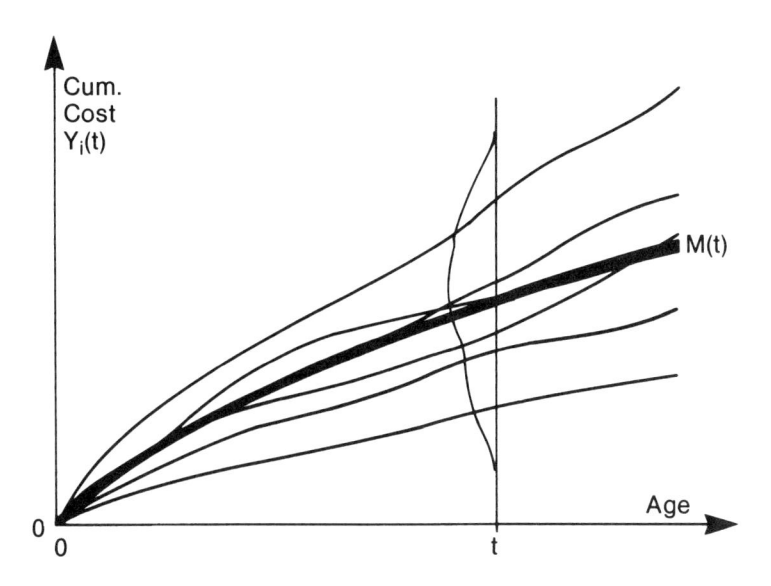

Figure 2.3. *Population of continuous cumulative cost histories.*

of that population unit. Thus the population "value" for each population unit is not a single number but is a *curve*. As before, any measure of the "value" of a unit may be used, for example, uptime of a system. Also, such cumulative history functions may decrease, for example, water taken from a reservoir.

Types of events. Such history functions may include all types of events, selected types of events, or a single type of event. The choice of what to include depends on the application and the desired information. In practice, one may analyze a data set any number of times using different choices of the types of events and their values. Different choices for repair data, for example, may yield engineering, management, or accounting information.

Termination. In some applications, the histories of some units terminate before the final age of interest. For example, some patients in a medical study may die before the study ends. Also, for example, a heat pump may be retired from service without running to a later age of interest. Such histories may not be randomly censored. The model and theory in this book must be modified to handle such termination. Wang, Qin, and Chiang (2001) provide a model and analysis for such nonrandom censoring (also called informative censoring).

2.3 Population Model and Its MCF

Purpose. This section presents the nonparametric population model, population distributions, and population MCF.

Model. The *nonparametric model* for a population of units is the population of the cumulative history functions of all units. That is, the model is a population of curves. Such a population of *continuous* curves appears in Figure 2.3. A population model for the cumulative *number* of discrete events is called a counting process. It is the population of all staircase functions, like that in Figure 2.1 for transmission repairs. If plotted on a figure like Figure 2.1, such a population of staircase functions would coincide so much that the figure would be unreadable. Similarly, a population model for the cumulative *cost* or *value* of discrete events is the population of all staircase functions like the one in Figure 2.2 for cumulative costs of fan motor repairs.

Stochastic process. This nonparametric model is a simple representation of a stochastic process. In stochastic process theory, any observed cumulative history function is called a realization of the process, a sample path, and an outcome.

Data. Age data on sample history functions may have right, left, or gap censoring. Censoring is a property of the data and its collection, not the model. As stated before, the population history functions in the model have no censoring.

Population. To use this model, one must have a clearly defined physical target population of units. This practical issue is discussed in detail in Chapter 1.

Not assumed. Note that this nonparametric model does not use a parametric stochastic process to generate such curves and entails no assumptions about the history functions. Anything can physically produce the functions. The few assumptions for this model and the lack of the usual assumptions for other models are spelled out in detail in Chapters 3 and 4.

Continuous distribution. At any age t, the corresponding values of the population cumulative history functions have a distribution. These values are the heights of the points where the history functions pass through the vertical line at age t (see Figure 2.3). This distribution differs at different ages. For a large population of *costs* or *values*, this distribution might be regarded as continuous. Figure 2.3 shows a continuous distribution density. There the density should be perpendicular to the page, but it is depicted on its side for better viewing.

Discrete distribution. Figure 2.4 shows five population staircase history functions for the cumulative *number* of events. In this figure, it is difficult to see the individual histories, since they overlap; the continuous histories in Figure 2.3 are easier to see. In Figure 2.5, a large population of such functions at age t has a discrete distribution of the cumulative numbers of events, which has integer values $0, 1, 2, 3, \ldots$. These values are the heights of all staircase functions at age t. At age t, a fraction of the population has accumulated zero recurrences, another fraction of the population has accumulated one recurrence, another fraction has accumulated two recurrences, etc. The lengths of the bars in Figure 2.5 are proportional to these fractions. The bars should be perpendicular to the page, but they are laid flat for easy viewing. Thus these bars depict the discrete population distribution of the cumulative number of recurrences at age t. This distribution differs at different ages.

MCF. At any age t, the corresponding distribution of the number or cost of events has a mean $M(t)$. This mean as a function of t is called the *mean cumulative function* (MCF) (see Figures 2.3 and 2.5). For costs, it is called the MCF for *costs* (of events). For the number of events, it is called the MCF for the *number* of events. This function can be regarded as the "mean curve," as it is the pointwise average of all population curves passing through the vertical line at each age t. Usually, this population mean curve is a step function with many small steps, one for each event in the population. For many applications, the mean curve is regarded as continuous. In most applications, $M(t)$ is an increasing function of age t. This mean curve yields most of the information sought from recurrence data, as described in section 2.4. The rest of this book is mainly concerned with estimating $M(t)$ and obtaining information from the estimate. For those acquainted with stochastic counting processes, the MCF for the number of events is usually denoted by $\Lambda(t)$.

Rate. Often for count data, the derivative

$$m(t) \equiv \frac{dM(t)}{dt}$$

is assumed to exist. It is called the instantaneous population *recurrence rate* or *intensity function*. It is called "instantaneous" because it depends on the age t. The word "instantaneous" is often omitted but implied. $m(t)$ is expressed in the number of events per month

2.3. Population Model and Its MCF 27

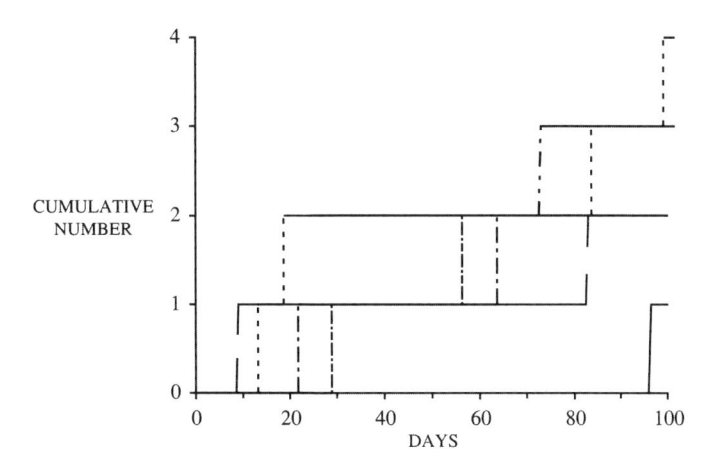

Figure 2.4. *Population histories for the cumulative number of recurrences.*

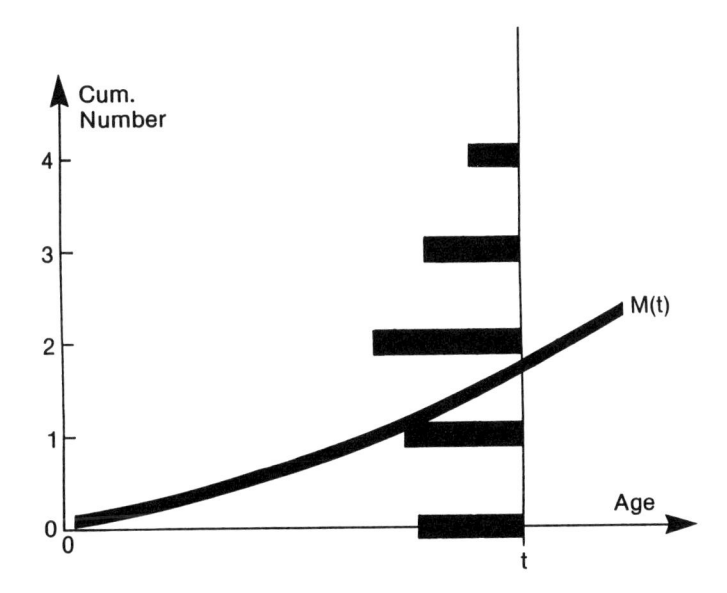

Figure 2.5. *Discrete distribution of cumulative number of recurrences at age t.*

(hour, year, mile, etc.) per population unit, for example, the number of fan motor repairs per month per heat pump. Figure 2.6 shows three MCFs with (a) increasing, (b) constant, and (c) decreasing recurrence rates. In reliability work, the behavior of the recurrence rate is used to make decisions on product burn-in, preventive replacement, and retirement, as explained later. In stochastic process theory, the recurrence rate for the number of events is usually denoted by $\lambda(t)$.

Failure rate. In reliability work, some call $m(t)$ the "failure rate," which causes confusion with the failure rate (hazard function) associated with a life distribution for a nonrepaired item (usually a component). The hazard function has an entirely different definition, meaning, and use. This confusion is greatest for renewal data on a system that is replaced when it fails; such data are described by renewal theory (Chapter 8). Ascher

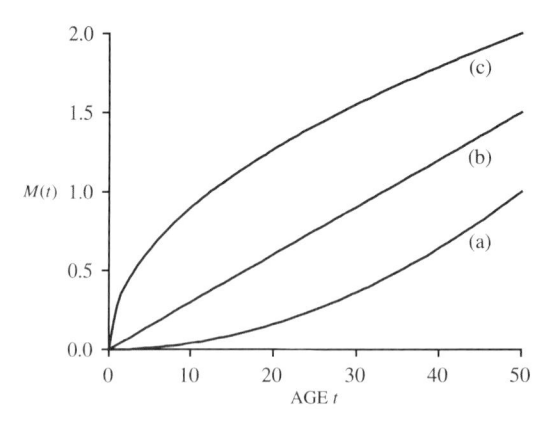

Figure 2.6. *MCFs with* (a) *increasing,* (b) *constant, and* (c) *decreasing recurrence rates.*

and Feingold (1984) and Ascher (2003) carefully distinguish the recurrence rate of system repairs and the failure rate (hazard function) for the life of nonrepaired items.

Life data. In work with life data (survival data), there is a *single* event for each unit, namely, end of life. To model life data, one uses a cumulative distribution function $F(t)$ for the population fraction "dead" by age t. The distribution probability density is the derivative $f(t) = dF(t)/dt$. The instantaneous failure rate (hazard function) for a life distribution is $h(t) \equiv f(t)/[1 - F(t)]$. The cumulative hazard function is $H(t) \equiv \int_0^t h(x)dx$, and $F(t) = 1 - \exp[-H(t)]$. For the purposes of this book, these life distribution concepts are quite unrelated to the MCF $M(t)$ and the recurrence rate $m(t)$ for repeated events data.

Cost rate. Similarly, for cost or value data, the derivative $m(t)$ is the population *cost rate* (or value rate). It has the dimensions of, say, dollars per month per population unit.

2.4 Information Sought

Purpose. This section shows how the MCF is used to obtain information from recurrence data. The need for such information was described in Chapter 1. Where plots of a sample MCF appear below, imagine a smooth curve drawn through the data points as an estimate $M^*(t)$ of the population $M(t)$. Chapters 3 and 5 explain how to obtain such plots.

Estimate $M(t')$. In many applications, one seeks an estimate of $M(t')$, the population MCF at a specified age t'. For example, for the car transmissions, the manufacturer wanted an estimate of the mean cumulative number of repairs during design life, that is, at $t' = 24,000$ test miles (132,000 customer miles). Chapter 3 gives this estimate as 0.31 transmission repairs per car; equivalently, $R/100 = 31$ repairs per hundred cars or 31%.

Rate behavior. One often seeks to know the behavior of the population recurrence rate $m(t)$, which is estimated from the slope of a sample MCF. Is it constant over time? Does it increase or decrease with the age of the population? Figure 2.6 shows MCFs with increasing, constant, and decreasing recurrence rates. In reliability work, knowledge of recurrence rate behavior is used to make decisions on product burn-in, preventive replacement, and retirement.

Turbines. Figure 2.7 displays the sample MCF for 21 marine turbines versus their operating hours. This line printer plot shows the number of repairs (points) in a rectangle on the plot. Also, numbers greater than 9 are represented by letters ($A = 10$, $B = 11$, etc.).

2.4. Information Sought

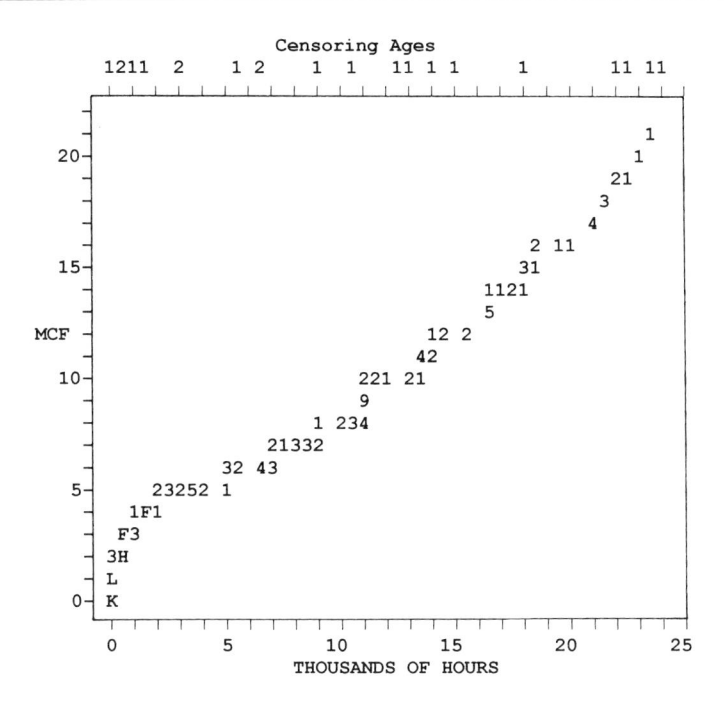

Figure 2.7. *"Bathtub" curve for the MCF of turbines.*

The top line of the plot (labeled "Censoring Ages") displays the 21 censoring ages—useful information. This plot shows a "bathtub" recurrence rate (that is, derivative) for repairs. The initial rate (slope) is high and decreases; this behavior is typical of startup problems of complex systems and is called "infant mortality." The rate decreases up to 3000 hours and then slowly increases thereafter; the increasing rate after 3000 hours is called "wearout behavior," as it is often related to physical wearout of products. Many reliability texts state that such a bathtub failure rate is typical of products. In the author's experience, it describes about 10% of products. Note that no other applications in this book have a bathtub failure rate. A bathtub failure rate is more typical of large systems with many failure modes.

Burn-in. Some products are subjected to a factory burn-in. The units are run and repaired as needed until the population (instantaneous) recurrence rate decreases to some desired value m'. An estimate of the suitable length t' of burn-in is obtained from the sample MCF as shown in Figure 2.8. A straight line segment with slope m' is moved parallel to itself until it is tangent to the MCF. The corresponding age at the tangent point is the desired t', as shown in Figure 2.8.

Blood analyzers. Figure 2.9 shows a plot of the sample MCF from data collected from blood analyzers during a burn-in period. The purpose of the plot was to assess the effect of burn-in and to determine a suitable length of burn-in. The manufacturer sought a recurrence rate m' below one repair per 10,000 actuations (blood samples), which is the slope of the line in the plot. The plot shows important features. First, repair data before 2000 actuations obviously were not provided. Second, the repairs accumulated linearly between 2000 and 5000 actuations; this corresponds to a constant rate. Third, beyond 5000 actuations, the repairs accumulated linearly but at a lower (constant) rate. The clear change of slope at 5000 actuations is obviously a human artifact; physical failure processes produce smooth curves. When questioned, the test engineers acknowledged that they had used different

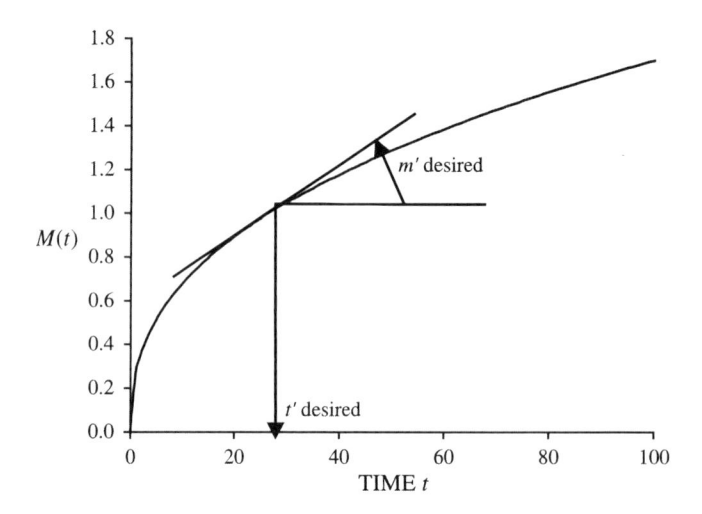

Figure 2.8. *Method to estimate the desired length t' of burn-in.*

test procedures and definitions of failure before and after 5000 actuations. This shows how plots yield important and unexpected insights, which numerical methods would not. The test procedure and failure definition after 5000 actuations correspond to the customer's. (Subsequent data analyses, not shown here, employed a consistent definition of failure.) The constant rate before and after 5000 cycles suggests that burn-in beyond 2000 actuations does not reduce the recurrence rate for repairs. The rate (slope) beyond 5000 actuations is below the desired rate of $m' = \frac{1}{10,000}$.

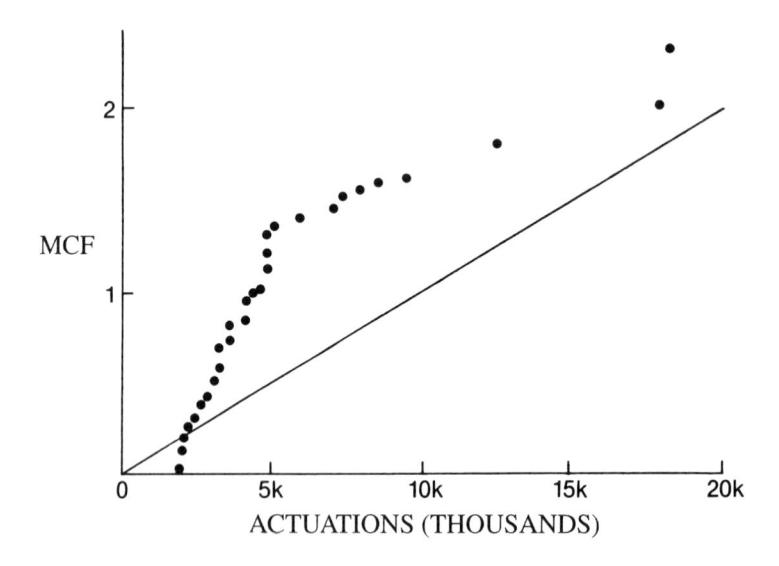

Figure 2.9. *Blood analyzer sample MCF.*

Availability. A typical cumulative history function for uptime of a unit appears in Figure 1.7. The availability of a fleet of units can be seen from a plot of the MCF for uptime versus scheduled running time. Such a fleet MCF looks much like Figure 2.8. The slope of a

2.4. Information Sought

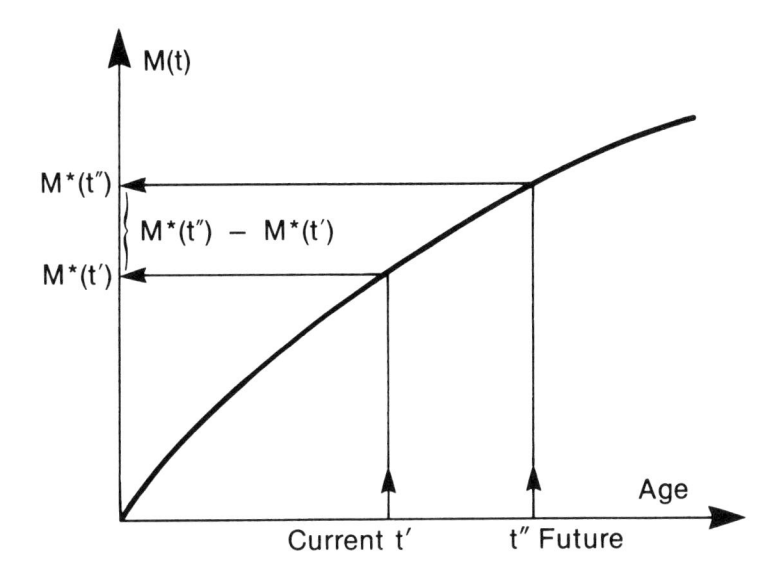

Figure 2.10. *Depiction of prediction of future cost or number.*

line tangent to the MCF at a particular age t' is the instantaneous availability of fleet units at that age, that is, the expected fraction of units that are up at that age. A straight line segment from the time origin to the tangent point has a slope that equals the average fleet availability over that time period.

Prediction. The MCF can be used to predict the total future number or cost of events in a population of units in a future period as follows. For a particular unit, suppose that the unit's ages at the start and end of the period are t' and t'', respectively. To predict cost, use the sample MCF $M^*(t)$ for cost to obtain $M^*(t')$ and $M^*(t'')$. Their difference $M^*(t'') - M^*(t')$ is a prediction of the expected incremental future cost for that unit. This difference is shown graphically in Figure 2.10. The sum of such predictions for all units is the prediction of the total population cost in the future period. Of course, t' and t'' differ from one unit to another. Also, units that enter the population during the future period can be included in the calculation; for them, $t' = 0$. This calculation encounters a problem if the age t'' of some unit exceeds the greatest (censoring) age for which $M^*(t)$ has been estimated. Then one must fit a suitable curve (by eye or by mathematics) to the plot and extrapolate the curve, possibly with much error, to greater ages as needed. Prediction of the future number of events similarly employs $M^*(t)$ for the number of events.

Ultimate $M(\infty)$. In most applications, the MCF increases without limit as the population ages. In others, the MCF flattens out and approaches an asymptote denoted by $M(\infty)$, called the ultimate MCF value. This value is of interest in some applications. For example, for the births data, this is the ultimate mean number of children born to statisticians (and their mates). This mean number should exceed 2.1 children, if the parents are to replace themselves. Figure 2.11 shows such flattened sample MCFs for the number of births to male and female statisticians; there the ultimate number of children is about 1.4. Another example would be a study of a population of savings accounts, which eventually will all close. There the bank wants to know how much profit it will ultimately make on such accounts; this profit will determine the tax writeoff on the accounts, as they are a depleting asset. Similarly, in a study of treatment costs of a population of medical patients with a particular disease, the ultimate (total) mean cost of treatment per patient is of interest.

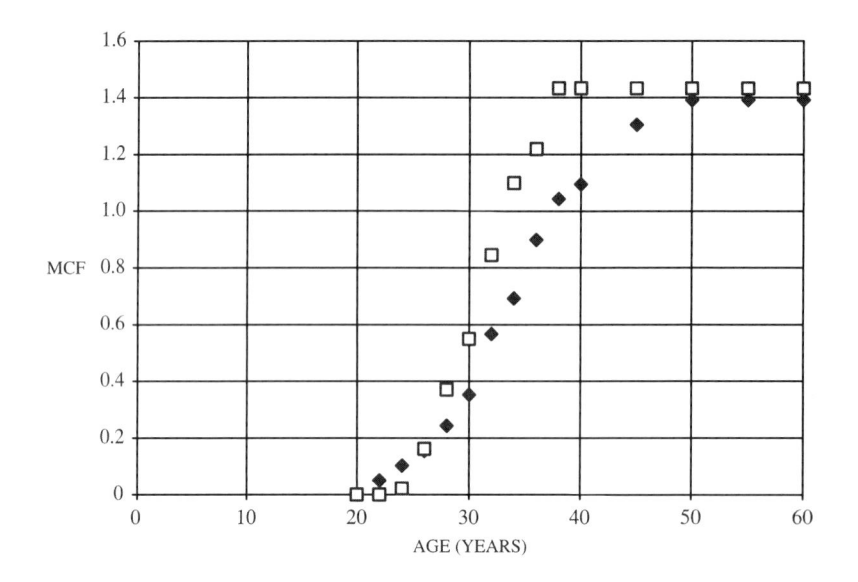

Figure 2.11. *MCFs for births to male ♦ and female ☐ statisticians.*

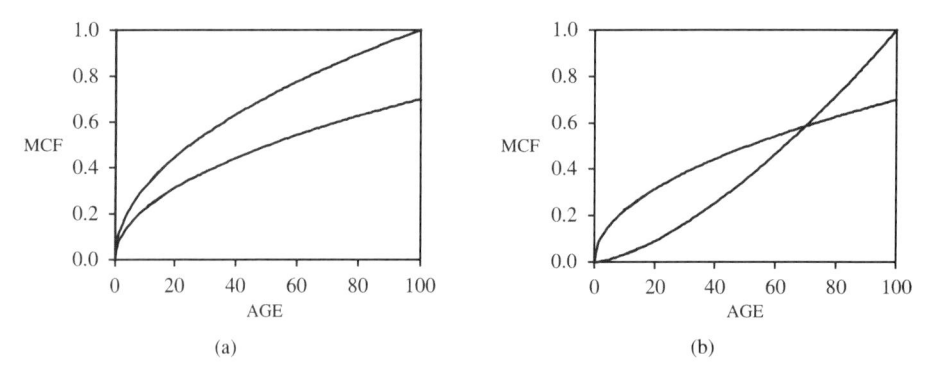

Figure 2.12. *Comparison of MCFs.*

Retirement. An optimum retirement age of a system (or fleet) can be determined from its sample (mean) cumulative cost function. Such a cost function would include initial purchase price and scrap value (negative cost) as well as repair and other maintenance costs. How to determine the optimum retirement age is not explained here.

Comparisons. MCFs can be used to compare different populations. For example, Morin (2002) compares the purchasing behavior of groups of Internet shopping customers receiving different sales promotions. Similarly, populations of manufactured products can differ with respect to design, operation, maintenance, environment, application, etc. Then one wishes to know if population MCFs differ, in order to gain information on how to improve a product. If one sample function is below another, as in Figure 2.12(a), the cost or number of recurrences for that population is lower. If such functions cross, as in Figure 2.12(b), then one must decide whether low cost or number is more desirable at early ages or later ages. Of course, it is essential that the difference between sample MCFs be convincing (statistically significant) and also big enough for practical purposes (practically significant). Otherwise, taking action is not warranted. Statistical confidence limits (Chapter 4) help one judge if differences are convincing.

Table 2.1. *Proschan air conditioner hours at repair (** denotes major overhaul).*

7907	7908	7909	7910	7911	7912	7913	7914	7915	7916	7917	8044	8045
194	413	90	74	55	23	97	50	359	50	130	487	102
209	427	100	131	375	284	148	94	368	304	623	505	311
250	485	160	179	431	371	159	196	380	309		605	325
279	522	346	208	535	378	163	268	650	592		612	382
312	622	407	710	755	498	304	290	1253	627		710	436
493	687	456	722	994	512	322	329	1256	639		715	468
	696	470	792	1041	574	464	332	1360			800	535
	865	494	813	1287	621	532	347	1362			891	594
	1312	550	842	1463	846	609	544	1800			934	728
	1496	570	1228	1645	917	689	732				1164	880
	1532	649	1287	1678	1163	690	811				1167	907
	1733	733	1314	**	1184	706	899				1297	921
	1851	777	**	1693	1226	812	945					1151
	**	836	1467	1797	1246	1018	950					1217
	1885	865	1493	1832	1251	1100	955					1278
	1916	983	1819		1263	1154	991					1312
	1934	1008			1383	1185	1013					
	1952	1164			1394	1401	1152					
	2019	1474			1397	1447	1362					
	2076	1550			1411	1558	1459					
	2138	1576			1482	1597	1489					
	2145	1620			1493	1660	1512					
	2167	1643			1507	1678	1525					
	2201	1705			1518	1869	1539					
		**			1534	1887						
		1835			1624	2050						
		2043			1625	2074						
		2113			1641							
		2214			1693							
		2422			1788							

Other information. The preceding types of information all come from the MCF. While not used in the author's experience, other information from such data can be imagined. For example, for a particular age, one might be interested in features of the distribution of cost or number of failures, for example, its standard deviation and percentiles. Also, one might want to assess whether the population has independent increments, which simplifies parametric modeling and data analysis.

Problems

2.1. Your application. For a set of your data, describe in detail the information you seek from the data and how that information will contribute to practical decisions.

2.2. Proschan data. Proschan (2000) published the failure ages in Table 2.1 in running hours for 13 airplane air conditioner systems, where ** denotes a major overhaul. Proschan modeled each air conditioner as a Poisson process (Chapter 8), each with a different recurrence rate. That is, times between failures for an air conditioner are statistically independent and from the same exponential distribution. He states that his model would be used in predicting reliability, scheduling maintenance, and ordering spare parts, but he does not explain how.

(a) Make a time-event display of the data. Comment on what the display shows.

(b) Plot the 13 history functions on separate plots but all to the same scale. Under Proschan's model, each plot should be relatively straight. Comment on the suitability of his model for each air conditioner. Do different means for the systems seem suitable?

(c) Describe how to assess whether times between failures in a system are statistically independent.

Chapter 3
MCF Estimates for Exact Age Data

3.1 Introduction

Purpose. This chapter shows how to calculate a nonparametric estimate of the population MCF from various types of data and how to plot and interpret the estimate. The MCF plot provides most of the information sought from recurrence data, and it is as important as probability plots are for life and other data. Necessary background is found in Chapters 1 and 2.

Advantages of plots. Data plots have the following advantages and should always be used, whether analytic methods are used or not:

- Plots are simple, quick to make, and easy to interpret. Common statistical packages make such plots, and many plots are easy to make with a spreadsheet.

- They provide an estimate of the population MCF and other quantities of interest.

- They allow one to assess how well a proposed parametric model fits the data.

- They reveal peculiar data, which often yield valuable insights.

- They help convince others of conclusions based on the plots or on analytic methods.

- They reveal unsought insights into the data. The MCF plot for the blood analyzers (Chapter 2) and the defrost control (Problem 5.2) are examples of this. Analytic methods rarely reveal anything not specifically calculated.

Disadvantages. Plots have the following disadvantages:

- The accuracy of estimates from plots is unknown, although some experienced analysts can subjectively judge accuracy. It is best to also use analytic methods to calculate and plot confidence limits, which are objective. Inexperienced analysts tend to think estimates are more accurate than they actually are.

- Graphical comparisons of data sets may be inconclusive unless sets differ greatly. Such comparisons are more accurate if aided by confidence limits or hypothesis tests.

- Plots cannot be used to determine appropriate sample sizes. Analytic theory provides this.

In most work, it is best to use a combination of plots and analytic methods.

Motivation. Figure 3.1 motivates the MCF estimate $M^*(t)$ for the population $M(t)$ at age t when histories are uncensored. Suppose that the N cumulative history functions for cost (or recurrences) in the figure are a simple random sample from the population. Also, suppose that the N observed cumulative costs at age t are $Y_1, Y_2, \ldots, Y_N$. At age t, the natural estimate of the population mean $M(t)$ is the sample mean $\bar{Y} = Y_1 + Y_2 + \cdots + Y_N/N$. When the histories are censored, the MCF estimate presented below must be used.

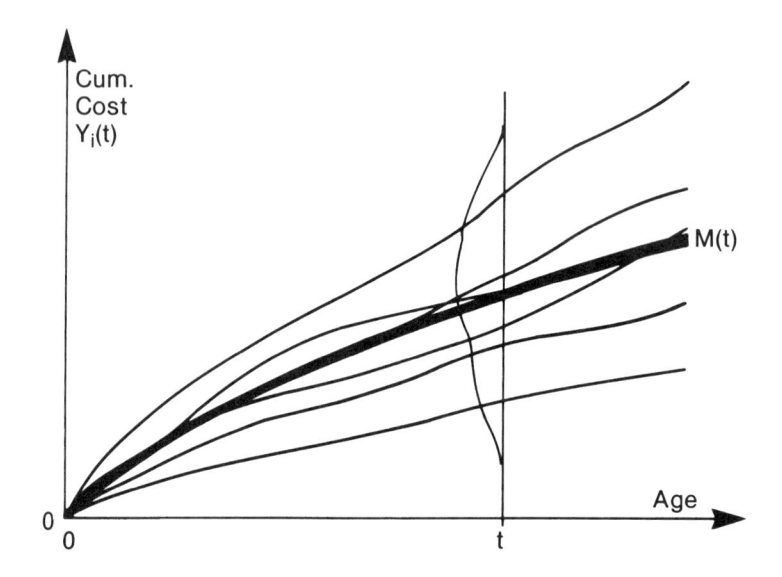

Figure 3.1. *A sample of uncensored cumulative histories.*

Population and sample. Chapter 1 describes the issues in specifying the target population. In this chapter, we deal with the analysis of sample data. The sample must be chosen appropriately, so that it is representative of the target population. Statistical theory requires that the sample be truly random. This issue of random sampling is discussed in detail in section 3.5. In practice, many samples are not random and yield estimates with unknown biases.

Chapter overview. The following sections of this chapter deal with estimating the MCF for various types of recurrence data, which are described in Chapter 1, and the practical and theoretical issues:

3.2. *MCF for number from exact age data with right censoring.*

3.3. *MCF for cost from exact age data with right censoring.*

3.4. *MCF from exact age data with left censoring and gaps.*

3.5. *MCF from continuous history function data.*

3.6. *Practical and theoretical issues.*

Problems.

Table 3.1. *Automatic transmission MCF calculations.*

1. Mileage	2. Number r observed	3. Mean number 1/r	4. MCF
28	34	1/34 = 0.03	0.03
48	34	1/34 = 0.03	0.03 + 0.03 = 0.06
375	34	1/34 = 0.03	0.03 + 0.06 = 0.09
530	34	1/34 = 0.03	0.03 + 0.09 = 0.12
1388	34	1/34 = 0.03	0.03 + 0.12 = 0.15
1440	34	1/34 = 0.03	0.03 + 0.15 = 0.18
5094	34	1/34 = 0.03	0.03 + 0.18 = 0.21
7068	34	1/34 = 0.03	0.03 + 0.21 = 0.24
8250	34	1/34 = 0.03	0.03 + 0.24 = 0.27
13809+	33		
14235+	32		
14281+	32		
17844+	30		
17955+	29		
18228+	28		
18676+	27		
19175+	26		
19250	26	1/26 = 0.04	0.04 + 0.27 = 0.31
19321+	25		
19403+	24		
19507+	23		
19607+	22		
20425+	21		
20890+	20		
20997+	19		
21133+	18		
21144+	17		
21237+	16		
21401+	15		
21585+	14		
21691+	13		
21762+	12		
21876+	11		
21888+	10		
21974+	9		
22149+	8		
22486+	7		
22637+	6		
22854+	5		
23520+	4		
24177+	3		
25660+	2		
26744+	1		
29834+	0		

3.2 MCF for Number from Exact Age Data with Right Censoring

Purpose. This section shows how to calculate, plot, and interpret the MCF estimate for the number of recurrences from exact age data with right censoring, the most common form of recurrence data. The calculation is illustrated with the transmission repair data and the tumor recurrence data, which are both discussed in Chapter 1.

Transmission data. Table 3.1 displays data on 10 repairs on $N = 34$ cars with automatic transmissions, originally appearing in Table 1.1. Information sought from the data includes (1) the mean (cumulative) number of repairs per car by 24,000 test miles (or 132,000 customer miles, the design life) and (2) whether the population recurrence rate for repairs increases or decreases as the population ages.

MCF estimate. The following steps yield a nonparametric estimate $M^*(t)$ of the population MCF $M(t)$ for the *number* of recurrences from a sample of N units. The display of the transmission data in Figure 3.2(a) provides a visualization of these steps. $M^*(t)$ is an unbiased estimator for $M(t)$. This estimate is also called the "sample MCF." It is similar to Nelson's (1982, 2000a) nonparametric estimate (hazard plot) of the sample cumulative hazard function of a life distribution of items that fail once. Programs that do these calculations are reviewed later.

(a) Order ages. List all sample recurrence and censoring ages in order from smallest to largest, as shown in column 1 of Table 3.1. Denote each censoring age with a +. If a recurrence age for a unit is the same as its censoring age, put the recurrence first. If two or more units have a common recurrence or censoring age (ties), list them in a suitable order, possibly random. In the transmission data, there are no ties. In Figure 3.2(a), one can imagine the vertical line moving from age 0 to the right and passing through the successive recurrence and censoring times in the order listed in Table 3.1.

(b) Number at risk. For each sample age, write in column 2 the number r of remaining units ("at risk" of recurrence) at that age as follows. For the earliest age, write $r = N - 1$ if it is censored. Otherwise, write $r = N$ if it is a recurrence. Then proceed down column 2, writing the preceding number r for each recurrence age or writing the preceding number minus 1 for each censoring age. That is, the remaining observed number r decreases by 1 at each censoring age. For the last age, which is always censored, $r = 0$. In Figure 3.2(a), if one visualizes the vertical line located at a sample age, then the number of history lines that it passes through is the number of sample units remaining (at risk) at that age. For example, in Figure 3.2(a), the vertical line is at the last repair (19,250 miles), and the number remaining at that age is 26.

(c) Mean number. For each recurrence, calculate its observed incremental "mean number of recurrences per unit" at that age as $1/r$ in column 3. That is, one out of the r units that passed through that age had a recurrence then. For example, for the repair at 19,250 miles, the increment is $1/26 = 0.04$ in column 3. If a unit has 2, 3, etc., multiple recurrences at an age, use $2/r$, $3/r$, etc. For a censoring age, the observed mean number of recurrences is 0, shown as a blank in column 3. The censoring ages determine the r values of the recurrences; thus the calculation takes into account the censoring ages.

(d) MCF. In column 4, calculate the value of the sample MCF $M^*(t)$ at each recurrence by summing the preceding increments as follows. For the earliest recurrence age, its MCF value is its mean number of recurrences in column 3, namely, 0.03 in Table 3.1. For each successive recurrence age, its MCF value is its incremental mean number of recurrences (column 3) plus the preceding MCF value (column 4). For example, at 19,250 miles, the MCF value is $0.04 + 0.27 = 0.31$. No MCF value is calculated for censoring ages, but they are taken into account, as they determine the number at risk for each recurrence.

(e) Plot. Choose a graph grid, usually a linear grid, but a log-log grid may be useful. Choose the horizontal age scale to include the range of the age data, and choose the vertical MCF scale to include the MCF values. On the graph, plot each recurrence's MCF value (column 4) against its age (column 1), as in Figure 3.2(b), produced by a line printer. Censoring ages are not plotted. This plot displays the nonparametric estimate $M^*(t)$, also called the *sample MCF*. The sample MCF extends only to the last censoring age. Therefore, it is useful to display that age (and all other censoring ages) on the age axis to show the range of the data

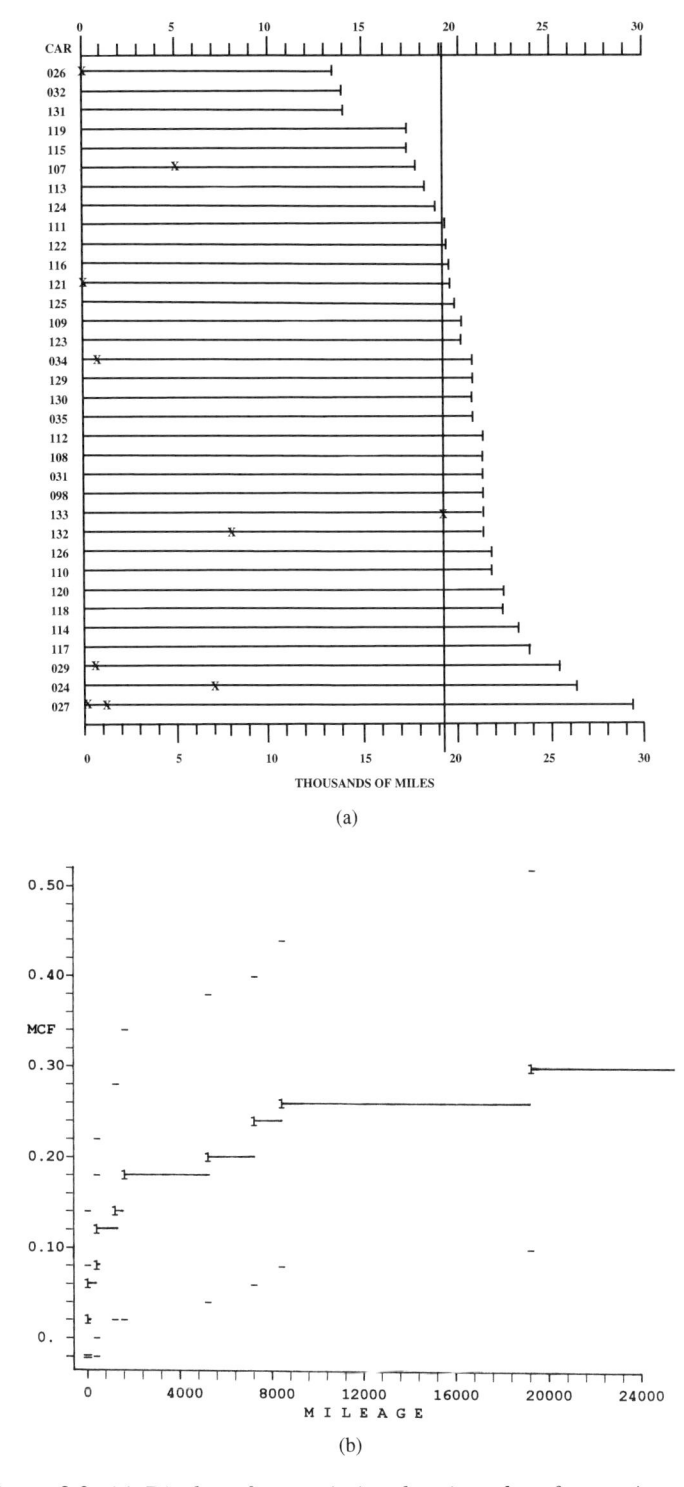

3.2. MCF for Number from Exact Age Data with Right Censoring

Figure 3.2. (a) *Display of transmission data in order of censoring age.* (b) *Automatic transmission MCF with 95% confidence limits.*

and MCF estimate. Interpretations of such plots are given with the various applications.

Staircase estimate. This nonparametric estimate $M^*(t)$ involves no assumptions about the population $M(t)$ or the process that generated the unit histories. Consequently, (1) this nonparametric estimate is a staircase function, which is flat between recurrence ages. Usually the flat portions distract the eye from the data points and are not plotted. Also, (2) the estimate extends to the longest censoring age. For example, the transmission estimate extends to 29,834 miles. Finally, (3) this estimate is unbiased, as shown in section 3.5. Usually the true MCF of a large population is regarded as a smooth curve. Then one imagines a smooth curve passing through the plotted points. Some may wish to fit a curve by hand or by mathematical means, such as least squares. Least squares confidence limits are not valid. The few assumptions underlying this estimate appear in section 3.6. It is essential to understand and verify them to get good results with the estimate. It is common, but poor, practice to merely state assumptions without assessing them.

MCF value. Often one wants an estimate of the population MCF at a specified age. This value is read directly from the staircase estimate or from a curve through the plotted points. For example, from Figure 3.1, the staircase estimate of the MCF at 24,000 test miles is 0.31 recurrences per car (during design life), answering a basic question. This estimate can also be read from Table 3.1. Equivalently, the $R/100$ (repairs per hundred cars) is 31 or 31%.

Recurrence rate. The derivative of such a curve (imagined or fitted) estimates the population recurrence rate $m(t)$. If the derivative increases with age, the population recurrence rate increases as units age. If the derivative decreases, the population recurrence rate decreases with age. The behavior of a product's recurrence rate is used to determine burn-in, overhaul, and retirement policies. In Figure 3.2, the recurrence rate (derivative) decreases as the transmission population ages, the answer to a basic question. Thus this population has an infant mortality behavior. This is typical of preproduction data, which are used to identify and correct production startup problems for new designs.

Software. The preceding estimate is easy to calculate with a pocket calculator or spreadsheet and plot by hand or computer. Programs that calculate this estimate for $M(t)$ from exact age data with right censoring and plot it with approximate confidence limits are as follows:

- MCFLIM of Nelson and Doganaksoy (1989). This was used to obtain Figure 3.2. Two-sided 95% confidence limits appear above and below each data point as dashes.

- The Reliability Procedure in the SAS/QC® software of the SAS Institute (1999, pp. 947–951).

- The JMP software of the SAS Institute (2000, pp. 23–27 and 92–95).

- SPLIDA features developed by Meeker and Escobar (2002) for S-PLUS of Insightful (2001).

- A program developed for General Motors by Robinson (1995).

- The ReliaSoft (2000a, b) Weibull++ software has a recurrence data add-on.

Size effect. There are populations with units of different sizes, a covariate. For example, underground power cables of a utility have different lengths. Repair data on such units can be plotted as described here, if the method is modified as described by Nelson (1982, p. 18), based on the series-system model.

Tumor data. Table 1.2 of Chapter 1 displays tumor data for the $N = 48$ patients on placebo treatment with a total of 87 recurrences. Table 3.2 shows part of the data ordered by

3.2. MCF for Number from Exact Age Data with Right Censoring

age from smallest to largest. Also, Table 3.2 shows part of the MCF estimate calculated by the SAS Institute (1999) Reliability Procedure. The tumor study was conducted to compare how three treatments affect tumor recurrence. Also, the investigators sought to understand how the rate of tumor recurrence behaves over time, that is, over the course of the disease. This knowledge aids in scheduling periodic examinations of patients. These data are analyzed here for the following reasons: (1) They are from another field, medicine; (2) the simple constant tumor recurrence rate is unlike that of other applications in this book; (3) the data show the output of another computer program. Here age data are recorded to the nearest month, and there are many age ties. These interval data are treated as exact ages here; this is often satisfactory when the age intervals are small compared to the time periods of interest.

Table 3.2. *Tumor data and SAS MCF calculations: The Reliability Procedure.*

```
                The RELIABILITY Procedure

                    Repair Data Summary
                                            Group
    Input Data Set          WORK.BLADDER    Placebo
    Total No. of Events          135
    No.of Units/Censoring Ages    48
    Number of Events              87

                    Repair Data Analysis

          Sample    Standard      95% Conf.Limits      Unit
   Age     MCF       Error       Lower      Upper       ID
   0.        .         .            .          .         1
   1.      0.021     0.021       -0.020      0.063      18
   1.        .         .            .          .         2
   2.      0.043     0.030       -0.016      0.102      48
   2.      0.065     0.037       -0.007      0.136      46
   2.      0.086     0.042        0.005      0.168      25
   2.      0.108     0.046        0.018      0.199      19
   3.      0.130     0.050        0.032      0.228      44
   3.      0.152     0.053        0.047      0.256      38
   3.      0.173     0.056        0.063      0.284      36
   3.      0.195     0.059        0.080      0.311      26
   3.      0.217     0.061        0.097      0.337      16
   3.      0.239     0.063        0.114      0.363      14
   3.      0.260     0.065        0.132      0.388      13
   4.        .         .            .          .         3
   5.      0.283     0.067        0.151      0.414       9
   5.      0.305     0.069        0.170      0.439      47
              Lines of output omitted here.
  46.      2.462     0.432        1.615      3.310      44
  48.        .         .            .          .        40
  49.      2.587     0.475        1.657      3.517      48
  49.      2.712     0.539        1.655      3.770      46
  49.        .         .            .          .        41
  51.      2.855     0.556        1.766      3.944      44
  51.        .         .            .          .        42
  52.      3.022     0.652        1.744      4.299      46
  53.      3.188     0.692        1.833      4.544      44
  53.        .         .            .          .        44
  53.        .         .            .          .        43
  59.        .         .            .          .        45
  61.        .         .            .          .        46
  64.        .         .            .          .        48
  64.        .         .            .          .        47
```

Chapter 5 describes more suitable analyses for interval age data.

Tumor MCF. Figure 3.3 shows the SAS plot of the placebo MCF, which increases linearly, starting at the origin. Thus the population recurrence rate (the derivative) is constant, the simplest possible behavior. This MCF and that for the Thiotepa treatment (Problem 3.2) are compared in Chapter 7. Figure 3.3 includes approximate 95% confidence limits, which are presented in Chapter 4. Also, SAS displays the sample censoring ages in a strip at the top of the plot, which helps one judge how many units went through any age.

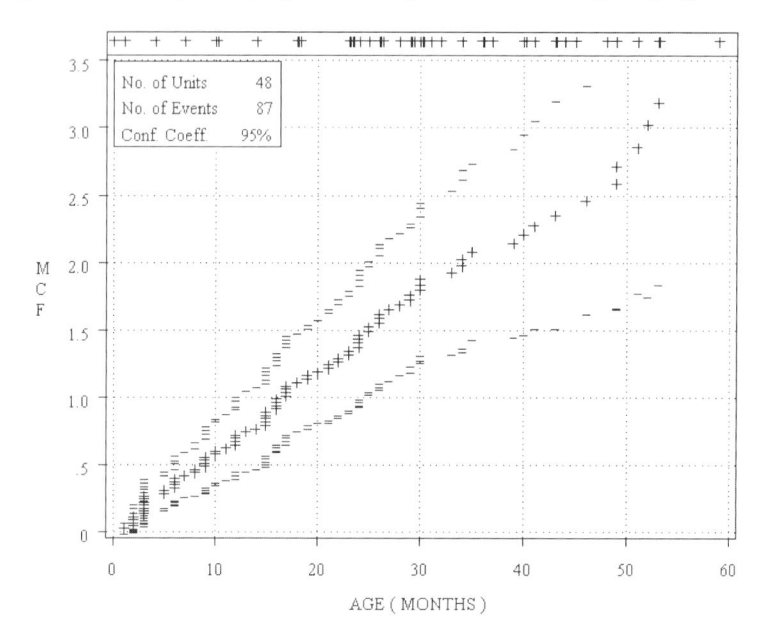

Figure 3.3. *Sample MCF of placebo data.*

Interpretation. The plot is essentially a straight line through the origin. Thus the population recurrence rate is constant, and patients do not get better or worse. Other analyses show that the rate differs from patient to patient. The plot intersects the horizontal 1.0 recurrence line at 17 months, which can be regarded as a typical time between recurrences, since the recurrence rate is constant.

Accuracy. If mistakenly regarded as a regression plot of independent observations, the MCF in Figure 3.3 looks quite accurate with 87 data points, one for each recurrence. On the contrary, there are just 48 independent patients, and the MCF is less accurate than it appears. Thus confidence limits (Chapter 4) are needed to properly judge its accuracy. Moreover, correct confidence limits look nothing like regression limits.

Censoring ages unknown. The MCF estimate here and elsewhere uses the censoring ages of all units in the sample. In some applications, censoring ages are unknown. For example, in automotive applications, the vehicle mileage is included in each failure report. However, the current mileages of all vehicles in the fleet may not be known. In such situations, it is necessary to approximate the distribution of censoring ages. Four common ways to do this follow. First, as described by Vasan et al. (2000), the heavy truck industry collects data on and publishes the distribution of annual mileage for various classes of trucks; such a distribution and the known starting dates of the units can be used to estimate their current mileages. Second, one can use production information and estimated dates of installation to approximate the censoring age distribution of units. Third, one can take a small random

3.3 MCF for Cost from Exact Age Data with Right Censoring

sample from the population and use their current ages to estimate the distribution of censoring ages, as described by Suzuki (1985) for life data. Fourth, as a last resort, those with experience can subjectively estimate the distribution of censoring ages. Hu, Lawless, and Suzuki (1998) deal with missing censoring ages for life data.

Purpose. This section shows how to calculate, plot, and interpret the MCF estimate for cost or value data from exact age data with right censoring, the most common form of cost data.

Cost data. The cost data on fan motor repairs in Table 1.3 of Chapter 1 are used here to explain how to calculate the sample MCF for *cost* (or value) data. These data on $N = 119$ heat pumps appear below in Table 3.3. The service contract provider wanted to know how such costs accumulate over time. The provider used this information to price service contracts, to predict costs, and to decide which brands of heat pump to exclude from service contracts.

MCF for cost. The following steps yield a plot of a nonparametric estimate $M^*(t)$ for the population MCF $M(t)$ for *cost*. $M^*(t)$ is an unbiased estimator for $M(t)$.

(a) Order ages. As before, list all sample recurrence and censoring ages in order from smallest to largest, as in column 1 of Table 3.3. Denote each censoring age with a +. For each recurrence, put its cost in column 2 next to its age. If a recurrence age of a unit equals its censoring age, put the recurrence age first. If two or more different units have a common age (recurrence or censoring), put the ties together in the list in any suitable order, say, random order. For example, there are five units with age 3631 days. The order of ties may have a slight effect on the MCF estimate and confidence limits.

(b) Number at risk. In column 3, write the observed number r of units at risk at each sample age as follows. Suppose that there are N sample units. If the earliest age is a *censoring* age, write $r = N - 1$. If the earliest age is a *recurrence* age, write $r = N$. In Table 3.3, the earliest age (141 days) is a repair, and $r = N = 119$ there. Proceed down column 3, writing the preceding number r for each recurrence age and writing the preceding number minus 1 for each censoring age. That is, the number observed r decreases by 1 at each censoring age. The last age in column 3 is always a censoring age, and always its $r = 0$; Table 3.3 is truncated at $r = 73$ and does not show the final $r = 0$.

(c) Mean cost. Proceeding down column 4, calculate the observed "mean cost per unit" for each recurrence. This is the cost of the recurrence in column 2 divided by the number r of units observed at that age. For example, for the repair at 843 days, the cost is $110.20, and $r = 107$ units went through that age. Thus its observed mean cost per unit at that age is $110.20/107 = $1.03 in column 4. For any censoring age, there is no mean cost per unit. However, the censoring ages determine the r values of the recurrences and thus are taken into account.

(d) MCF. Proceeding down column 5, calculate the MCF estimate for cost for each recurrence as follows. For the smallest recurrence age, the MCF is its mean cost per unit. In Table 3.3, this value is $0.37. For each successive recurrence, its MCF is its mean cost per unit in column 4 plus the preceding MCF estimate in column 5. For example, for the repair at 1269 days in Table 3.3, the MCF estimate is $1.30 + $1.40 = $2.70. No MCF estimates are calculated for censoring ages.

(e) Plot. As before, choose a plotting grid, usually a linear grid, but a log-log grid may be useful. Choose the horizontal age scale to include the range of the age data, and choose

Table 3.3. *Calculation of the fan motor cost MCF.*

1. Days	2. Cost ($)	3. Number at risk r	4. Mean cost $/r$	5. MCF for cost	6. Rate $m = 1/r$	7. MCF for number
141	44.20	119	44.20/119 = 0.37	0.37	0.008	0.008
252+		118				
288+		117				
358+		116				
365+		115				
376+		114				
376+		113				
381+		112				
444+		111				
651+		110				
699+		109				
820+		108				
831+		107				
843	110.20	107	110.20/107 = 1.03	1.03 + 0.37 = 1.40	0.009	0.018
880+		106				
966+		105				
973+		104				
1057+		103				
1170+		102				
1200+		101				
1232+		100				
1269	130.20	100	130.20/100 = 1.30	1.30 + 1.40 = 2.70	0.010	0.028
1355+		99				
1381	150.40	99	150.40/99 = 1.52	1.52 + 2.70 = 4.22	0.010	0.038
1471	113.40	99	113.40/99 = 1.15	1.15 + 4.22 = 5.37	0.010	0.048
1567	151.90	99	151.90/99 = 1.53	1.53 + 5.37 = 6.90	0.010	0.058
1642	191.20	99	191.20/99 = 1.93	1.93 + 6.90 = 8.83	0.010	0.068
1646	36.00	99	36.00/99 = 0.36	0.36 + 8.83 = 9.20	0.010	0.078
1762+		98				
1869+		97				
1881+		96				
1908	158.50	96	158.50/96 = 1.65	1.65 + 9.20 = 10.85	0.010	0.089
1920+		95				
2110+		94				
2261	243.80	94	243.80/94 = 2.59	2.59 + 10.85 = 13.44	0.011	0.099
2273	189.50	94	189.50/94 = 2.02	2.02 + 13.44 = 15.46	0.011	0.110
2363	225.40	94	225.40/94 = 2.40	2.40 + 15.46 = 17.86	0.011	0.121
2419+		93				
2440	197.80	93	197.80/93 = 2.13	2.13 + 17.86 = 19.98	0.011	0.131
2562+		92				
2593	184.00	92	184.00/92 = 2.00	2.00 + 19.98 = 21.98	0.011	0.142
2615+		91				
2674	167.20	91	167.20/91 = 1.84	1.84 + 21.98 = 23.82	0.011	0.153
2710	149.00	91	149.00/91 = 1.64	1.64 + 23.82 = 25.46	0.011	0.164
2794+		90				
2815+		89				
2838	42.00	89	42.00/89 = 0.47	0.47 + 25.46 = 25.93	0.011	0.175
2946	255.70	89	255.70/89 = 2.87	2.87 + 25.93 = 28.80	0.011	0.187
2951	243.30	89	243.30/89 = 2.73	2.73 + 28.80 = 31.54	0.011	0.198
2986+		88				
3017+		87				
3087+		86				
3216+		85				
3291+		84				
3296	145.00	84	145.00/84 = 1.73	1.73 + 31.54 = 33.26	0.012	0.210
3307	208.00	84	208.00/84 = 2.48	2.48 + 33.26 = 35.74	0.012	0.222
3368	248.70	84	248.70/84 = 2.96	2.96 + 35.74 = 38.70	0.012	0.234
3391	256.90	84	256.90/84 = 3.06	3.06 + 38.70 = 41.76	0.012	0.246
3406+		83				
3440	305.50	83	305.50/83 = 3.68	3.68 + 41.76 = 45.44	0.012	0.258
3489	202.80	83	202.80/83 = 2.44	2.44 + 45.44 = 47.88	0.012	0.270
3490+		82				
3621+		81				
3631+		80				
3631+		79				
3631+		78				
3631+		77				
3631+		76				
3635	242.00	76	242.00/76 = 3.18	3.18 + 47.88 = 51.07	0.013	0.283
3639+		75				
3648+		74				
3652+		73				

(The 73 remaining units are all censored and slightly above 3652 days.)

3.3. MCF for Cost from Exact Age Data with Right Censoring

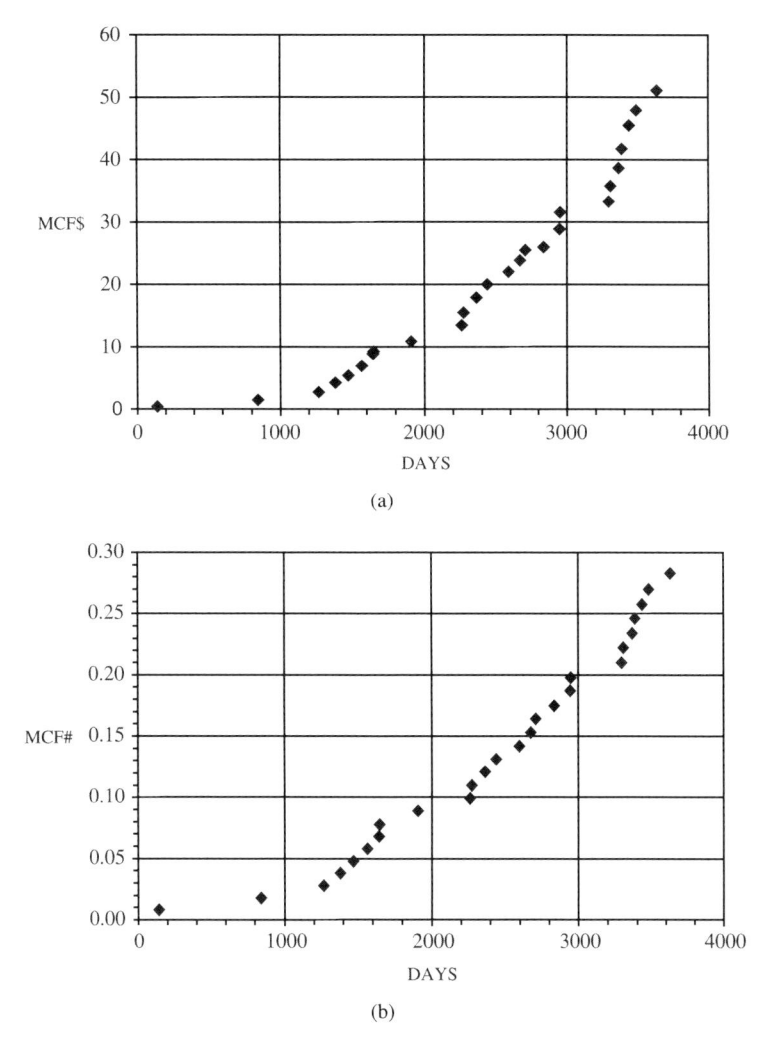

Figure 3.4. (a) *MCF of* cost *of fan motor repairs.* (b) *MCF of* number *of fan motor repairs.*

the vertical MCF scale to include the MCF values. For each recurrence, plot its MCF for cost in column 5 against its age in column 1, as in Figure 3.4(a). The resulting plot is the sample estimate $M^*(t)$ of the true population $M(t)$ for cost. Censoring ages are not plotted in the MCF. Flat portions of the sample MCF may be plotted if desired. As always, the sample MCF extends only to the last censoring age. Therefore, it is useful to display that age (and all other censoring ages) on the age axis to show the range of the data and MCF estimate. However, the $N = 119$ censoring ages do not appear in Figure 3.4(a).

Extensions. Columns 6 and 7 of Table 3.3 show the calculations for the sample MCF for the *number* of recurrences. There 1 is used in column 6 in place of the cost of a recurrence. The plot of this MCF appears in Figure 3.4(b).

Interpretation. Both plots show a relatively failure-free period, after which the recurrences and costs accumulate almost linearly. However, the cost plot is slightly curved upward. The probable cause of this curvature is price inflation, as the data extend over 10 years.

Software. The preceding MCF estimate for costs is easy to calculate with a pocket calculator or spreadsheet and plot by hand or computer. Excel generated Table 3.3. All the software packages described above calculate the sample MCF for *costs* and its approximate confidence limits and plot them for exact age data with right censoring.

3.4 MCF from Exact Age Data with Left Censoring and Gaps

Purpose. This section shows how to calculate, plot, and interpret the MCF estimate for the number of recurrences from exact age data with left and right censoring and gaps. Such data are less common in practice.

Compressor data. The MCF calculation is illustrated with the compressor data in Table 1.4 of Chapter 1. Table 3.4 displays the data for 28 compressor replacements on $N = 1322$ air conditioners. Information sought from the data includes (1) the MCF and (2) whether the population replacement rate increases or decreases as the population ages.

MCF estimate. The following steps yield a nonparametric estimate $M^*(t)$ of the population MCF $M(t)$ for the *number* of recurrences from a sample of N units.

(a) **Order ages.** List all I sample recurrence and left and right censoring ages t_i in order from smallest, t_1, to largest, t_I, as shown in column 1 of Table 3.4. Put tied ages in a suitable order, possibly random. Many of the air conditioners have the same censoring ages, but each unit could have distinct censoring ages.

(b) **Number entering.** In column 2, for each left censored age t_i, write the number of units N_i that *enter* the sample at that age. For example, at age 0, 149 units from Building H and 195 units from Building K enter the sample and are then at risk. Similarly, at age 1.00 years, 458 units from Building E enter the sample and are observed at risk. Similarly, for each right censored age t_i, write *minus* the number of units that *leave* the sample at that age. For example, at age 4.14 years, -195 units from Building H leave the sample and no longer were observed at risk. Column 2 has no entry for a recurrence age t_i since a recurrence does not change the observed number at risk.

(c) **Number at risk.** In column 3, calculate the observed number r_i at risk at each age t_i as follows. In the first line, set $r_1 = N_1$, the first number of units to enter the sample. In Table 3.4, $r_1 = N_1 = 149$. Then proceed down column 3, setting $r_i = r_{i-1} + N_i$. For example, $r_2 = r_1 + N_2 = 149 + 195 = 344$. r_i decreases when units leave the sample, where there is a negative N_i. Always the last $r_I = 0$, as in the last line of Table 3.4.

(d) **Increment.** For each *recurrence*, calculate its observed "incremental mean number" of recurrences per unit (at risk) as $m_i = 1/r_i$ in column 4. That is, one out of the r_i units observed passing through that age t_i had a recurrence then. If a unit has 2, 3, etc., multiple recurrences at an age t_i, use $2/r_i$, $3/r_i$, etc. Some prefer to work with the percentage $m_i = 100/r_i$. For example, for the earliest recurrence at 0.17 years, $m_3 = 100/344 = 0.29\%$ in column 4 of the table. For a *censoring age*, $m_i = 0$; this appears as a blank in column 4.

(e) **MCF.** In column 5, calculate the MCF estimate $M_i^* = M^*(t_i)$ at each recurrence as follows. For the earliest recurrence age, its MCF value is its mean number of recurrences, namely, $M_i^* = m_i = 1/r_i$. In Table 3.4, this is $M_3^* = m_3 = 0.29$. For each successive recurrence age, its MCF value is its incremental mean number of recurrences (m_i in column 3) plus the preceding MCF value (M_{i-1}^* in column 5). That is, $M_i^* = m_i + M_{i-1}^*$. For example, at 5.56 years, the MCF value is $0.10 + 1.80 = 1.90\%$. No MCF value is calculated for censoring ages. Discrepancies in the last decimal place come from rounding results to two decimal places.

3.4. MCF from Exact Age Data with Left Censoring and Gaps

Table 3.4. *MCF calculations for left censored age data.*

1. Age t_i	2. Enter N_i	3. Number at risk r_i	4. Incremental % $m_i = 100/r_i$	5. Cumulative % $M_i^* = M_{i-1}^* + m_i$
0	+149	149		
0	+195	344		
0.17		344	0.29	0.29
0.17		344	0.29	0.58
1.00	+458	802		
1.34		802	0.12	0.71
2.17		802	0.12	0.83
2.58		802	0.12	0.96
2.59	+164	966		
3.30		966	0.10	1.06
3.65		966	0.10	1.16
4.14		966	0.10	1.27
4.14	−195	771		
4.45	+356	1127		
4.47		1127	0.09	1.35
4.47		1127	0.09	1.44
4.62		1127	0.09	1.53
4.62		1127	0.09	1.62
4.65		1127	0.09	1.71
4.79		1127	0.09	1.80
5.09	−149	978		
5.56		978	0.10	1.90
5.57		978	0.10	2.00
5.75		978	0.10	2.11
5.75		978	0.10	2.21
5.80		978	0.10	2.31
5.85		978	0.10	2.41
6.13		978	0.10	2.51
6.73		978	0.10	2.62
7.02		978	0.10	2.72
7.05	−356	622		
7.33	−458	164		
7.42		164	0.61	3.33
7.42		164	0.61	3.94
8.77		164	0.61	4.55
9.27		164	0.61	5.16
9.27		164	0.61	5.77
9.33	−164	0		

(f) Plot. Choose a graph grid, usually a linear grid, but a log-log grid may be useful. Choose the horizontal age scale to include the range of the age data, and choose the vertical MCF scale to include the MCF values. On the graph, plot each recurrence's MCF value M_i^* (column 4) against its age t_i (column 1), as in Figures 3.5(a) and (b). Figure 3.5(a) displays the right and left censoring ages on the bottom axis.

Staircase estimate. As before, this nonparametric estimate $M^*(t)$ is a staircase function, which is flat between recurrence ages. The flat portions need not be plotted. The estimate extends to the last censoring age, 9.33 years for the compressors. This estimate is unbiased, as shown in section 3.5. Usually the true MCF of a large population is regarded as a smooth curve. Then one imagines a smooth curve passing through the plotted points. Some analysts may wish to fit a curve by hand or by mathematical means, such as least squares;

48　　　　　　　　　　　　　　　　Chapter 3. MCF Estimates for Exact Age Data

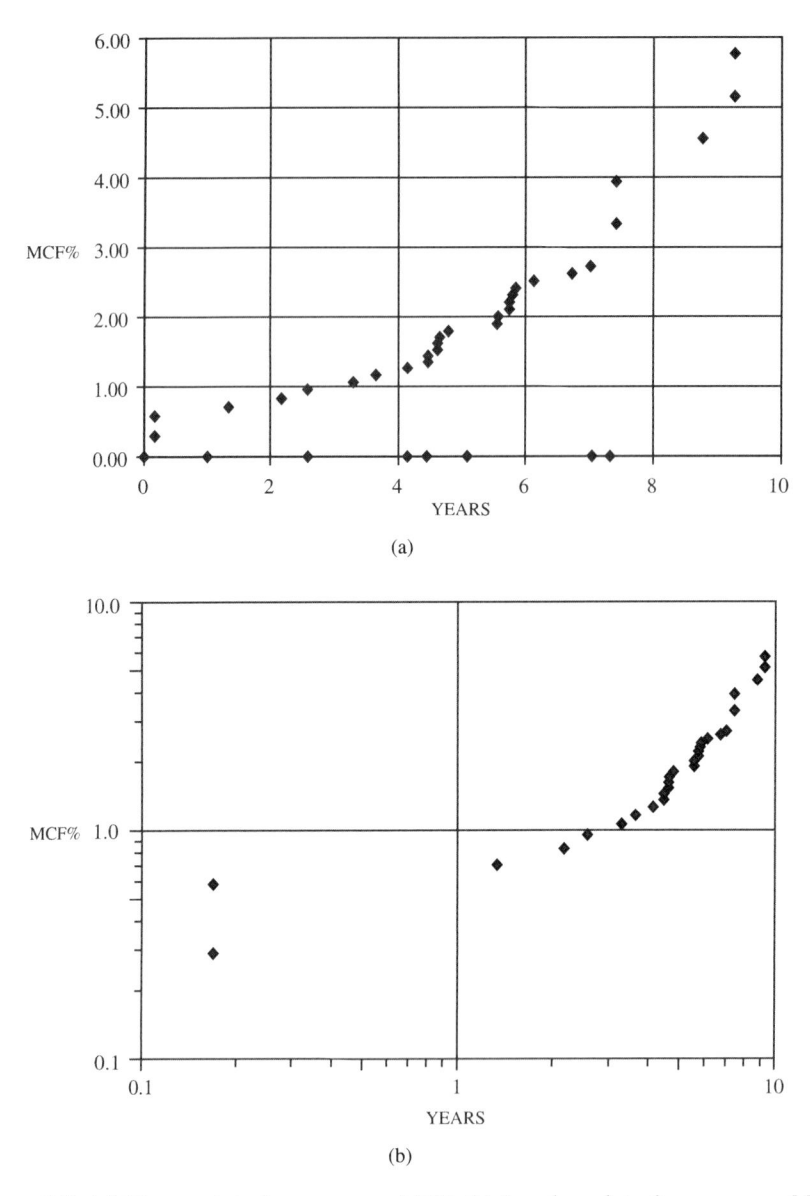

Figure 3.5. (a) *Linear plot of compressor MCF.* (b) *Log-log plot of compressor MCF.*

however, least squares confidence limits are not valid. The few assumptions underlying this estimate appear in section 3.6. It is essential to understand and verify them to get good results with the estimate.

Interpretation. The linear plot (Figure 3.5(a)) is curved upwards (increasing derivative). This indicates that such compressors have an increasing replacement rate. The log-log plot (Figure 3.5(b)) is also curved upwards. Renewal theory (Chapter 8) for such compressor replacement says that this log-log MCF plot is straight below 10% replacement if the compressor life distribution is Weibull. Thus this plot indicates that the compressor life distribution is not Weibull. Indeed the population is a mixture of subpopulations, with each building having a different life distribution.

3.5. MCF from Continuous History Function Data

Extensions. The preceding MCF estimate extends as follows:

- In the compressor data, the units enter and leave the sample in groups (buildings). More generally, individual units can enter and leave the sample one at a time.

- When a unit's history has gaps, the unit leaves the sample at the start of a gap and reenters the sample at the end of the gap. Then the number at risk at a particular age is still the number of units observed passing through that age.

- The estimate extends in the obvious way to cost or value data. For a recurrence with an observed value Y_i, the observed increment is $m_i = Y_i/r_i$, where r_i is the observed number at risk at that recurrence.

- The estimate extends to continuous history functions (section 3.5) in the obvious way.

Software. Available software does not calculate the MCF estimate for exact age data with left censoring and gaps. The calculations and plot are easy to carry out with a spreadsheet. Table 3.4 and Figures 3.5(a) and (b) were generated with Excel.

3.5 MCF from Continuous History Function Data

Purpose. This section shows how to estimate the population MCF when the sample units have an observed variable with continuous history functions and exact censoring ages. The section briefly reviews applications, describes typical data, presents the MCF estimate, motivates the estimate, describes extensions to other censoring, and explains how to get the estimate from existing software.

Applications. Section 1.5 of Chapter 1 describes various applications with continuous history functions. These include energy output of a power plant; uptime of a power plant, locomotive, or other system; and defect initiation and growth.

Data. Denote the sample size by N. Denote the continuous function of sample unit i by $Y_i(t)$, observed from time 0 to its censoring age t_i. Also, order the units by censoring age and number them backwards with reverse ranks so that $t_N < t_{N-1} < \cdots < t_2 < t_1$. Figure 3.6(a) depicts such a sample of five history functions. The censoring times are all assumed to be distinct here to simplify the discussion. The censoring ages divide the age scale into N intervals: $[0, t_N), [t_N, t_{N-1}), \ldots, [t_3, t_2), [t_2, t_1)$. The notation $[,)$ indicates that the left endpoint is included in the interval and the right endpoint is not, a reasonable convention.

MCF estimate. The following formula provides the MCF estimate $M^*(t)$. Motivation for the estimate appears later. Suppose that an age t of interest is in interval r; that is, $t_{r+1} \leq t < t_r$. Define $t_{N+1} = 0$. Then

$$M^*(t) = \frac{\{[Y_1(t_N) - Y_1(t_{N+1})] + [Y_2(t_N) - Y_2(t_{N+1})] + \cdots + [Y_N(t_N) - Y_N(t_{N+1})]\}}{N}$$
$$+ \frac{\{[Y_1(t_{N-1}) - Y_1(t_N)] + [Y_2(t_{N-1}) - Y_2(t_N)] + \cdots + [Y_{N-1}(t_{N-1}) - Y_{N-1}(t_N)]\}}{N-1}$$
$$\vdots$$
$$+ \frac{\{[Y_1(t_{r+1}) - Y_1(t_r)] + [Y_2(t_{r+1}) - Y_2(t_r)] + \cdots + [Y_{r+1}(t_{r+1}) - Y_{r+1}(t_r)]\}}{r+1}$$
$$+ \frac{\{[Y_1(t) - Y_1(t_{r+1})] + [Y_2(t) - Y_2(t_{r+1})] + \cdots + [Y_r(t) - Y_r(t_{r+1})]\}}{r}.$$

50   Chapter 3. MCF Estimates for Exact Age Data

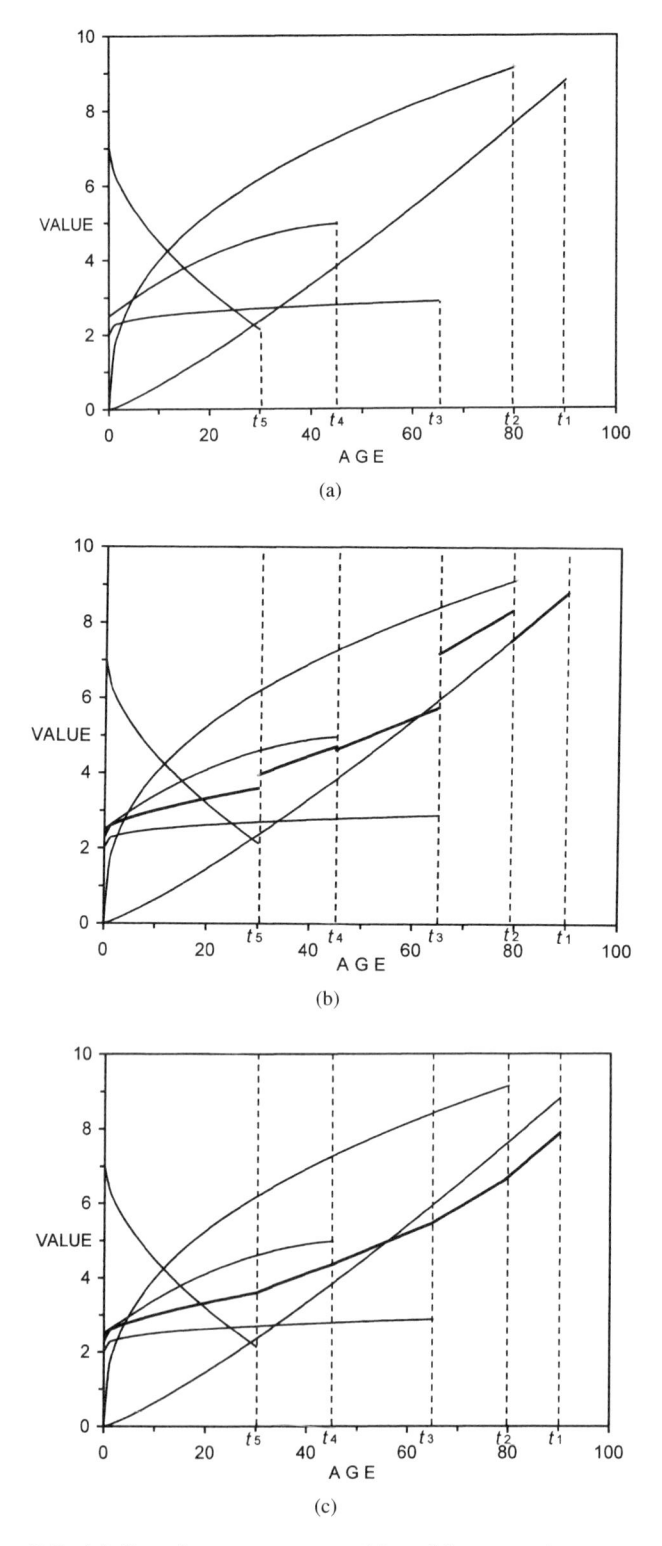

Figure 3.6. (a) *Sample age range partitioned by censoring ages t_i.* (b) *Average curves in each interval.* (c) *Linked average curves yield the sample MCF.*

Motivation. Figure 3.6(a) helps motivate this estimate. Note that a history function drops from the sample at each censoring age t_i. Also, note that the increment in $Y_i(t)$ for unit i in passing through interval j is $[Y_i(t_j) - Y_i(t_{j+1})]$. The first line of the estimate is the mean increment of the N sample units that passed through the earliest interval $[0, t_N)$, interval N. The second line is the mean increment of the $N - 1$ units that pass through interval $N - 1$. The last line is the mean increment of the r units that passed through $[t_{r+1}, t)$, part of interval r. The estimate $M^*(t)$ is just the sum of all these increments. Another view of this MCF estimate involves each interval's "mean curve," namely, the average of the sample history functions that pass through that interval, shown as the heavy curve segments in Figure 3.6(b). Then the sample "mean curve" $M^*(t)$ consists of the heavy mean curve segments, joined end-to-end by sliding them up or down, as shown in Figure 3.6(c).

Extensions. The preceding MCF estimate applies to any continuous cumulative history functions, including those that increase or decrease. Also, history functions may have (positive or negative) jumps at discrete ages and may be staircase functions. If they are all staircase functions, they correspond to discrete events, and the estimate above reduces to those in previous sections. The estimate above can be extended in the obvious way to continuous history functions with left censoring and gaps.

Software. There are no software packages that calculate the preceding MCF estimate for continuous history functions. The packages described previously can approximate the estimate by approximating the history functions with staircase functions as follows. First, divide the age scale from 0 to t_1 into a large number J of small intervals, with endpoints $0 < \alpha_j < \alpha_{j-1} < \cdots < \alpha_2 < \alpha_1 = t_1$. Then, for sample unit i at endpoint α_j, assign the increment $Y_{ij} = Y_i(\alpha_j) - Y_i(\alpha_{j+1})$. Thus the continuous sample history function $Y_i(t)$ is approximated with a staircase function with increment (jump) Y_{ij} at age α_j or at $\alpha_j + \alpha_{j+1}/2$, the midpoint. These staircase (discrete event) data can be put into the software packages, which will yield an approximate MCF estimate. The smaller the intervals, the better the approximation.

3.6 Practical and Theoretical Issues

Purpose. This section discusses the few assumptions and limitations of the MCF estimate and its theoretical properties. Properly dealing with all issues is essential to getting good results with the MCF estimate in practice. Assumptions are numbered in angled brackets, ⟨#⟩, and must be verified, not merely stated, in practice to ensure good results. Properties of estimates are numbered in curly brackets, {#}.

Population. As discussed in Chapter 1, ⟨1⟩ the target population is clearly specified and sampled. This assumption does *not* hold, for example, if prototype units are tested under laboratory conditions in order to predict the performance of subsequent production units in field service. However, engineers often must nevertheless base decisions on prototypes in the lab.

Random sample. It is assumed here that ⟨2⟩ the sample units are a simple random sample from the target population. This is a fundamental requirement in all statistical work. Strictly speaking, *simple random sampling* requires that every *sample* of size N units from the population have an equal chance of being selected. Often it is incorrectly stated that every *unit* must have an equal chance of being selected; many other types of random sampling have this property. Most statistical theory, including that for the MCF estimate here, requires simple random sampling. Such sampling can usually be achieved only with a complete list of the population and the use of random numbers to select the sample. Cochran (1977) discusses simple random sampling and other random sampling methods.

Biased sample. Rarely in engineering work or in other fields is a sample obtained by simple random sampling or by truly random sampling of any kind. Indeed, no sample in this book was truly random. Instead, practitioners must consider whether the sample is seriously biased and then analyze it anyway as though it were truly random. Such *sample bias* causes the statistical theory to be inaccurate. This is a reality that must be recognized and minimized. The defrost control application (Chapter 5) is an example of a seriously biased nonrandom sample, which was self-selected.

Random censoring. The MCF estimate rests on another basic statistical assumption, namely, that ⟨3⟩ the cumulative history functions of all N sample units are statistically independent of their censoring ages. Equivalently, ⟨3'⟩ the N censoring ages are randomly assigned to the N sample cost histories. Such random censoring is also called *noninformative*. This assumption of *random censoring* does not hold, for example, if a unit is retired early because it has had a costly repair history.

Nonrandom censoring. The following example has nonrandom (informative) censoring that occurs in practice. In the automotive industry, units are grouped by month of production, and each monthly group is analyzed separately. Figure 3.7 shows the sample MCFs for the number of repairs of a component from production month 1 and from month 2. Month 1 has been followed for four months to date, and its sample MCF is a straight line. Month 2 has been followed for three months to date, and its sample MCF is a straight line with a lower recurrence rate. The dotted line continuation shows the still-unobserved MCF from age 3.00 to 4.00 months. (Straight lines are used to simplify this discussion.) As is typical, the first production month has a higher repair rate. Suppose that both groups have the same number of units. Then the heavy line is the sample MCF from four months of data for month 1 and three months of data for month 2. This estimate is biased high from age 3.00 to 4.00 months. In contrast, if four months of data from month 2 were available, the correct sample MCF would be the dotted line. The problem here is that the censoring is

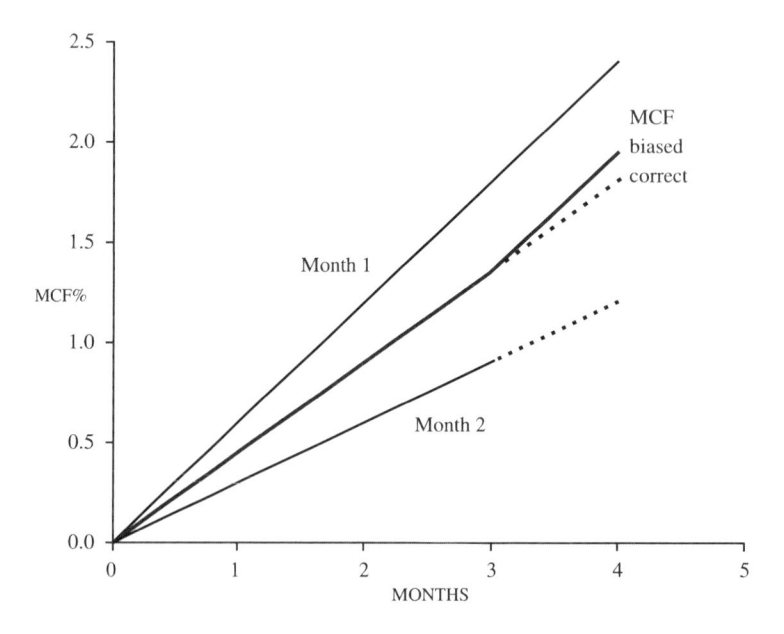

Figure 3.7. *A sample of uncensored cumulative histories.*

3.6. Practical and Theoretical Issues

biased. Specifically, only the high recurrence rate units from month 1 are observed through ages 3.00 to 4.00 months, and thus they bias the estimate high there. In such situations, one usually treats each group as a separate population and obtains a separate sample MCF for each, as is done in the automotive industry. Similarly, at the end of a medical or drug addiction study, a patient's history runs to the end of the study or ends earlier if the patient is lost from the study. Patient losses are usually informative and often indicate a poorer response to treatment. Morin's (2002) study of the purchasing behavior of Internet shopping customers also has possibly nonrandom loss of customers. Wang, Qin, and Chiang (2001) and their references present models and data analyses for dealing with informative censoring.

Reporting delays. Another form of nonrandom censoring arises in the automotive industry. Dealers report repairs when they wish, and some reports are quite delayed. This results in an underestimate of the recurrence rate. Kalbfleisch, Lawless, and Robinson (1991) correct this bias by estimating the distribution of reporting delays and compensating for them. Lawless (1995b) makes a similar correction for AIDS reporting and insurance claims.

Finite population. Engineers usually regard their fleet, a finite number of units in service, as their sample and population. Statisticians usually regard a fleet as a sample of units from some theoretical infinite population. Most statistical methods concern estimates and confidence limits for parameters of the theoretical infinite population. Such methods may not be suitable for a finite population. Engineers want statistical inferences about the random costs and numbers of repairs of their finite fleet. The MCF estimate is suitable whether the population is regarded as infinite or the same as the sample.

Unbiased. As noted above, {3} the MCF estimate $M^*(t)$ for $M(t)$ is unbiased. The following proof applies to a finite fleet, as well as to an infinite population. Here ⟨4⟩ it is assumed that the history function of each population unit extends through any age of interest up to the greatest sample censoring age τ. The average of the sample history functions is the uncensored sample MCF, and its expectation is the population MCF. It is assumed that ⟨5⟩ the population MCF exists (is finite) at any age t of interest up to the greatest censoring age. Next, it is assumed again that ⟨3'⟩ the N distinct (for simplicity) sample censoring ages are randomly assigned to the N history functions. This assumption ⟨3'⟩ is equivalent to ⟨3⟩. Each of the $N!$ random assignments has the same probability $1/N!$, and each such censored sample yields an MCF estimate from age 0 to τ. It can be shown that the average of these $N!$ MCF estimates from all possible censorings of the sample is the MCF of the uncensored sample over $[0, \tau)$. In turn, the expectation of the censored sample MCF is the expectation of the uncensored sample MCF, which is the population MCF. That is, the censored sample MCF is an unbiased estimate of the population MCF.

Mean exists. The existence of the estimate $M^*(t)$ requires that ⟨5⟩ $M(t)$ be finite at any age of interest. As a practical matter, for a real population, $M(t)$ always exists. However, $M(t)$ may not exist for certain mathematical models.

Ties. To keep the theory simple, we assume that ⟨6⟩ all recurrence ages are distinct from each other and from the censoring ages, which need not be distinct from each other. When there are tied ages, the MCF estimate depends on their order in the tabular calculation of the sample MCF.

Seasonal effect. In heat pumps, cars, and various other products, there may be a seasonal pattern of certain repairs. The estimate here does not try to take such a pattern into account.

Not assumed. Most modeling and analysis of recurrence data involve assumptions not required by the nonparametric estimate here. Ascher and Feingold (1984), Engelhardt (1995), and Rigdon and Basu (2000) present such models, analyses, and assumptions at length. The following are some common assumptions *not* used here:

1. A parametric stochastic process model, such as the nonhomogeneous Poisson process, produces the population history functions. Chapter 8 surveys some common parametric models.

2. The population units have statistically independent histories. The estimate here is valid if sample unit histories are correlated from interactions or common environments, common production periods, etc. However, simple random sampling is essential to being able to ignore such dependencies.

3. The population of history functions has independent increments. That is, accumulated costs or numbers of recurrences in any nonoverlapping age intervals are statistically independent. In many populations there is dependence. Namely, units with initially high (low) accumulations continue later to have high (low) accumulations. Also, the MCF estimate here is valid whether or not there is serial correlation (or cause and effect) in the histories of units.

4. The population MCF is a specified parametric curve. Here population MCF is nonparametric.

5. Different types of recurrences (e.g., product failure modes) are statistically independent. The MCF estimates of Chapter 6 are valid whether such types are independent or dependent. Also, it is valid for any choice of components, failure modes, or repairs included in the analysis.

6. The same stochastic process generates each unit history. The estimate here is valid for any mixture of subpopulations, for example, units under different conditions (mild and severe) or units with different designs or from different production periods. But censoring must be random, as noted above. Proschan (2000) discusses the issue of mixtures.

7. Unit histories are generated by a renewal process. That is, times between failures of a unit are independent and identically distributed. The compressors (section 3.4), for example, have a renewal process, as compressors are replaced, not repaired. Chapter 8 briefly surveys renewal theory. The MCF estimate here is valid whether or not times between failures are independent or identically distributed.

8. Repair restores a unit to like-new condition or like-old condition.

Of course, the MCF estimate here still applies if any of these assumptions hold. Indeed, the estimate is nonparametric and thus is always valid under its assumptions. However, this nonparametric estimate may be less accurate than another estimate that uses added (possibly parametric) assumptions, provided they are valid.

Parametric models. There are various parametric models for recurrence data. Ascher and Feingold (1984), Engelhardt (1995), and Rigdon and Basu (2000) present many such models. Selected parametric models are surveyed in Chapter 8. The simplest such model is the Poisson process. Used for product repairs, that model assumes that repairs randomly occur at a constant rate and are statistically independent of each other. Nelson (1982, Chapter 6, section 2) gives Poisson data analyses for (1) estimates and confidence limits for the recurrence rate and (2) predictions and prediction limits for future numbers of recurrences. Any parametric model has model error, which may be negligible or significant.

Problems

3.1. Manual transmissions. The preproduction proving ground test of the new car model included 14 cars with manual transmissions. The repair data on them appear in Table 3.5.

(a) Make displays of these manual transmission data like Figures 1.1(a) and (b). Comment on notable features of the plot.

(b) Calculate the estimate of the MCF function for the number of recurrences for this sample. Do this with a pocket calculator or use a spreadsheet.

(c) Plot the MCF estimate on Figure 3.2 with the automatic transmission MCF.

(d) Estimate the mean cumulative number of recurrences at 24,000 test miles. Compare this estimate at design life with that for automatic transmissions.

(e) Describe the recurrence rate behavior of manual transmissions.

(f) Compare the sample MCFs of the manual and automatic transmissions (Tables 3.1 and 3.5). Do they differ convincingly? Why?

Table 3.5. *Manual transmission repairs.*

Car	Mileage (+ latest)
025	27099+
028	21999+
030	11891 27583+
097	19966+
099	26146+
100	3648 13957 23193+
101	19823+
102	2890 22707+
103	2714 19275+
104	19803+
105	19630+
106	22056+
127	22940+
128	3240 7690 18965+

3.2. Thiotepa treatment. The bladder tumor study included 38 patients who received Thiotepa treatment and experienced a total of 45 recurrences. Their recurrence data are given in Table 3.6.

(a) Calculate the estimate of the MCF function for the number of recurrences for this sample. Use a computer program, pocket calculator, or spreadsheet.

(b) Plot this MCF estimate on Figure 3.3 with the placebo MCF.

(c) Describe the recurrence rate behavior of tumors under this treatment.

(d) Compare the sample MCFs of the placebo and Thiotepa treatments. Do they differ convincingly? Why?

(e) As the plots are relatively straight, estimate the recurrence rate for each treatment. As a percentage, by how much does the Thiotepa treatment reduce the recurrence rate?

3.3. Proschan data. Use the Proschan (2000) data on air conditioner systems from Problem 2.2 (Table 2.1).

Table 3.6. *Tumor recurrence data for Thiotepa treatment.*

ID	Months
49	1+
50	1+
51	5 5+
52	9+
53	10+
54	13+
55	3 14+
56	1 3 5 7 10 17+
57	18+
58	17 18+
59	2 19+
60	17 19 21+
61	22+
62	25+
63	25+
64	25+
65	6 12 13 26+
66	6 27+
67	2 29+
68	26 35 36+
69	38+
70	22 23 27 32 39+
71	4 16 23 27 33 36 37 39+
72	24 26 29 40 40+
73	41+
74	41+
75	1 27 43+
76	44+
77	2 20 23 27 38 44+
78	45+
79	2 46+
80	46+
81	49+
82	50+
83	4 24 47 50+
84	54+
85	38 54+
86	59+

(a) Calculate and plot the sample MCF. Under Proschan's model, the MCF should be a straight line. Is a straight line plausible? Why?

(b) Using a straight line through the data, estimate the population recurrence rate for repairs. This simple estimate is used to schedule repair work and manufacture of replacement parts.

(c) In view of Proschan's model with a different mean time between failures for each system, would you prefer to use a separate prediction of future repairs for each system or a pooled prediction based on the MCF? Why?

3.4. Valve seats. 41 locomotives running near Beijing required valve seat replacements due to dusty operating conditions. The data appear in Nelson (1995a). The SPLIDA features of Meeker and Escobar (2002) produced the replacement data plots in Figures 3.8 and 3.9,

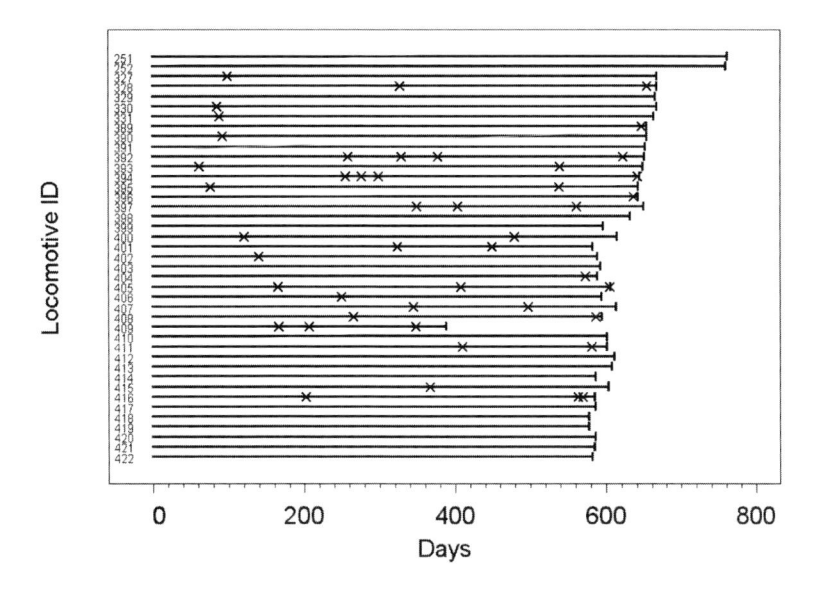

Figure 3.8. *Display of valve seat replacements on the 41 locomotives.*

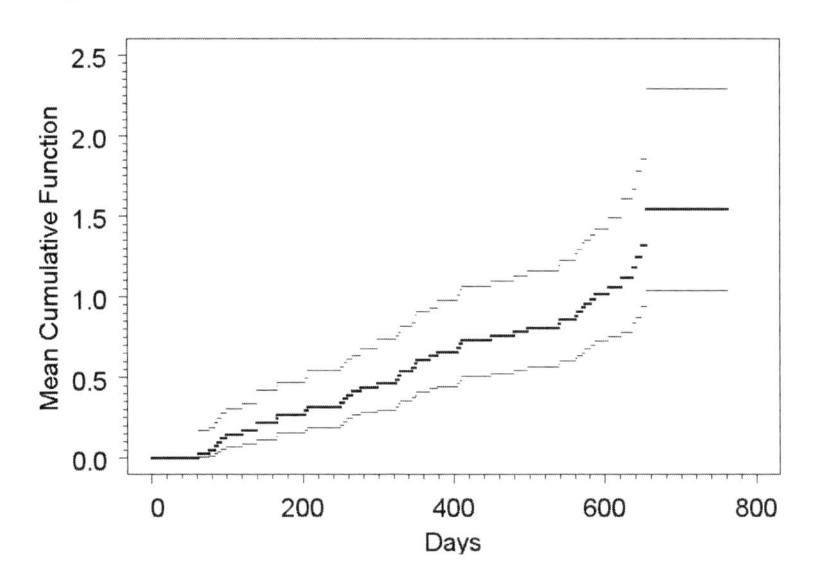

Figure 3.9. *MCF of locomotive valve seat replacements.*

which appear in Meeker and Escobar (1998, Chapter 16). They use a log transformation of the MCF estimate to get positive confidence limits, and thus their limits are not symmetric about the MCF estimate. The time-event display of the data shows that locomotive 409 has the fewest days in service. It was dropped into the water while being loaded on shipboard to go to China. Water removal and other cleanup delayed its start in service.

(a) The MCF plot is relatively straight after 90 days when valve seat inspections start. Estimate the replacement rate from the MCF plot.

(b) Predict the number of replacements per year for this fleet.

3.5. Circuit breakers. The laboratory test data in Table 3.7 come from 14 vacuum circuit breakers used in power systems. Failures are repaired.

(a) Calculate and plot the sample MCF.

(b) Describe and interpret the plot.

(c) The data have a relatively constant recurrence rate at early ages. Estimate it.

(d) Estimate the percentage of circuit breakers having a failure by 300 cycles.

Table 3.7. *Circuit breaker repair data.*

Breaker	Thousand cycles (+ censoring age)
1	20+
2	20+
3	44+
4	20+
5	21 25+
6	16 19 30+
7	16 18+
8	6 9 20+
9	3 11 19+
10	5 13 21+
11	11 20+
12	12+
13	18+
14	20+

3.6. Chronic granulomatous disease. Fleming and Harrington (1991, pp. 162–163 and 376–383) give data on recurrences of serious infections in patients with chronic granulomatous disease (CGD). Patients were randomly assigned to gamma interferon treatment or to placebo treatment.

(a) Calculate and plot the MCF for each treatment group, ignoring covariates.

(b) Compare the two MCFs. Do you think that the difference is convincing? Why?

(c) Use the covariates to divide the data into covariate groups, for example, using or not using corticosteroids, using or not using antibiotics, male or female, etc. Calculate and plot the MCF for each group. Note the differences that you think are convincing and why.

Chapter 4
MCF Confidence Limits for Exact Age Data

4.1 Introduction

Purpose. This chapter presents nonparametric approximate confidence limits for the MCF from exact age data with right censoring, the most common form of recurrence data. Such limits provide important information. First, they show the accuracy of the MCF estimate and thereby allow one to judge how much to rely on the estimate. Second, they can be used to compare MCF estimates of two or more samples, as described in Chapter 7. The background needed here is contained in Chapters 1, 2, and 3.

Overview. This chapter contains the following sections:

4.2 *Nelson's confidence limits*: This section presents correct approximate limits for the MCF that apply to the *number* of recurrences and also to the *cost* or other numerical *value* of recurrences. Such incremental costs and values may be negative. Also, these limits extend to continuous history functions.

4.3 *Naive confidence limits*: The naive limits for the MCF apply only to the *number* of recurrences. These approximate intervals are often appreciably shorter than the correct ones. Thus they make the MCF estimate appear more accurate than it really is. Also, approximate Poisson limits are presented; they are a special case of naive limits.

4.4 *Assumptions and theory*: This section describes assumptions for the approximate confidence limits. In practice, these assumptions must be evaluated to assess whether the confidence limits are satisfactory. This section also explains the theoretical properties of the limits.

Problems.

4.2 Nelson's Confidence Limits

Purpose. This section presents Nelson's (1995a) correct nonparametric approximate confidence limits for the population MCF from exact age data with right censoring on discrete events, the most common situation. This section motivates the limits, presents examples of applications (transmission and tumor data), describes properties of the limits, and reviews software that calculates and plots them. Section 4.4 explains the assumptions that must be

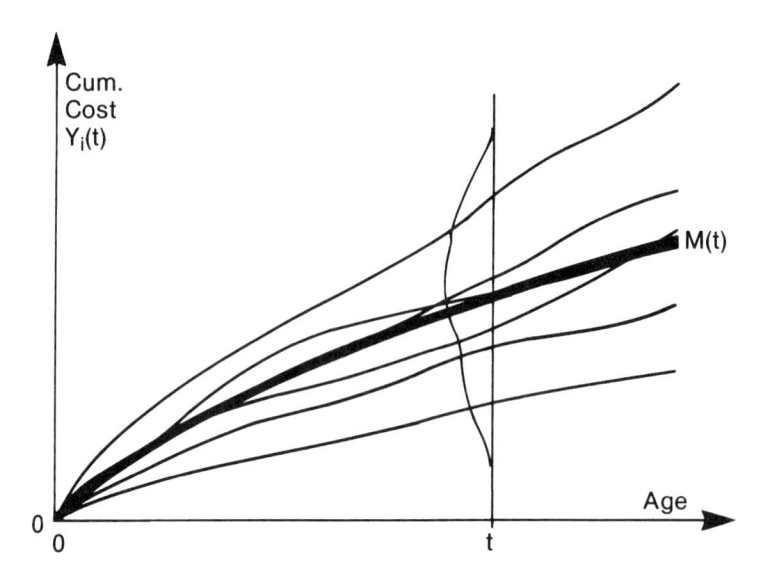

Figure 4.1. *Uncensored sample cumulative cost histories.*

verified to ensure that the limits are valid. Also, section 4.4 presents the laborious calculation of the limits, which requires a computer program, and their theoretical properties.

Motivation. Figure 4.1 motivates the variance of the MCF estimate $M^*(t)$ at age t. Suppose that the N cumulative history functions for cost in the figure are a simple random sample from an infinite population. These functions are usually staircases but are continuous here for ease of viewing. Also, suppose, for motivation, that all N sample histories are uncensored at age t, where the vertical distribution of cumulative cost appears. At this age, the estimate $M^*(t)$ of the population mean $M(t)$ at age t is the sample average $\bar{Y}$ of the N observed cumulative costs $Y_1, Y_2, \ldots, Y_N$ at age t. By the basic central limit theorem, for large N the sampling distribution of $M^*(t) = \bar{Y}$ is approximately normal with a true mean of $M(t)$ and true variance $V[M^*(t)] = \sigma_t^2/N$, where σ_t^2 is the variance of the depicted population distribution at age t. The estimate for $V[M^*(t)]$ is $v[M^*(t)] = s_t^2/N$, and s_t^2 is the usual (unbiased) sample variance of $Y_1, Y_2, \ldots, Y_N$. Thus two-sided normal approximate $C\%$ confidence limits for $M(t)$ are

$$\bar{Y} \pm K_C \left(\frac{s_t^2}{N}\right)^{1/2} = M^*(t) \pm K_C \{v[M^*(t)]\}^{1/2}, \qquad (4.1)$$

where K_C is the $\frac{1}{2}(100 + C)$ standard normal percentile. The theory in section 4.4 provides the true variance $V[M^*(t)]$ and its unbiased estimate $v[M^*(t)]$ when some sample history functions are censored before age t. This variance estimate used in the formula above yields approximate normal confidence limits for $M(t)$.

Transmissions. Table 1.1 presented data on repairs of automatic transmissions in a preproduction road test. Figure 3.2 displayed the sample MCF and approximate 95% confidence limits "−." Problem 3.1 presented similar data on repairs of manual transmissions, and Figure 4.2 below displays their sample MCF and approximate 95% confidence limits "−." The limit ˆ is off scale. Using these figures and the confidence limits, one can compare the manual and automatic transmissions with respect to recurrence rates for repairs. Clearly, the two sample MCFs do not differ statistically significantly over time, as the confidence

4.2. Nelson's Confidence Limits

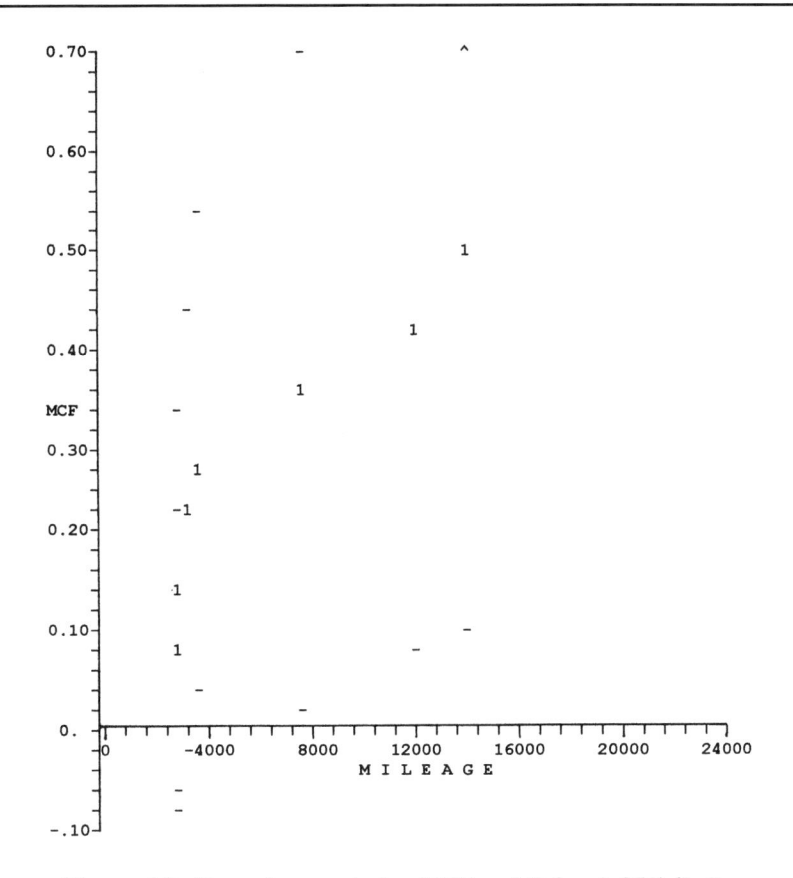

Figure 4.2. *Manual transmission MCF and Nelson's 95% limits.*

intervals are much wider than the difference of the two MCFs at any age. For the automatic transmissions, the nonparametric (staircase) estimate of the MCF at 24,000 test miles (design life) is 0.31, and the 95% limits are 0.09 and 0.51. These limits are quite wide due to the small sample, 34 cars, only four of which reached 24,000 test miles. Similarly, for the manual transmissions the staircase estimate of the MCF at 24,000 test miles is 0.50, and the 95% limits are 0.09 and 0.91. These limits are even wider due to the smaller sample, 14 cars, only three of which reached 24,000 test miles.

Tumors. Chapter 1 and Problem 3.2 presented data on recurrences of bladder tumors in patients under two treatments—placebo and Thiotepa. A main purpose was to compare these treatments. Figures 4.3(a) and (b) show SAS plots of the two sample MCFs with approximate 95% confidence limits. SAS data input and commands appear in Figure 4.3(c). The plots show that the treatments have an essentially constant recurrence rate. Also, the Thiotepa treatment has a lower recurrence rate than the placebo. The widths of the confidence intervals at the various ages are comparable to the difference between the two MCFs. Thus these figures do not clearly indicate whether these MCFs differ statistically significantly (convincingly). The method of comparison in Chapter 7 must be used to decide this. Note that the strip of +'s at the top of each plot displays the censoring ages of the sample.

Properties. Nelson's (1995a) confidence limits have the following properties:

- The limits are plotted only at the time of recurrences. Like the MCF estimate, the limits are staircase functions and are constant between events. In particular, a pair of limits

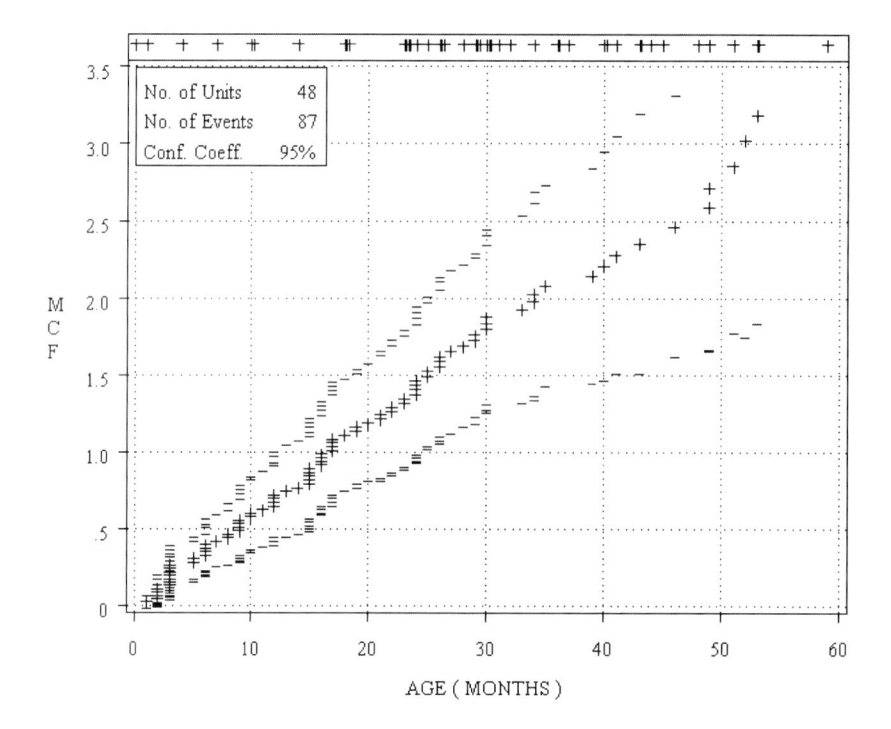

Figure 4.3(a). *Tumor MCF for placebo and Nelson's 95% limits.*

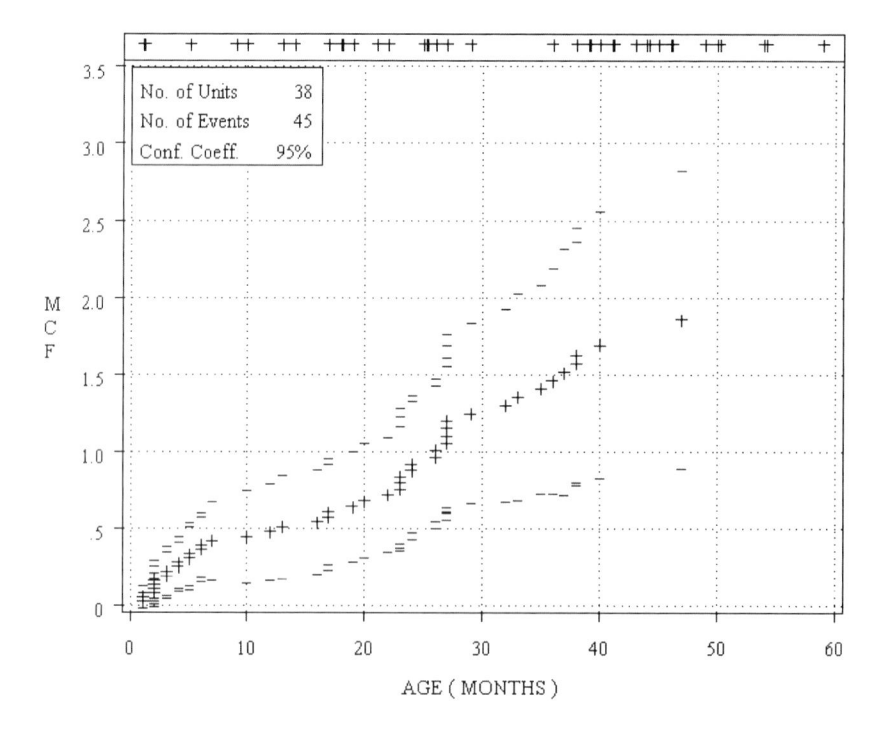

Figure 4.3(b). *Tumor MCF for Thiotepa and Nelson's 95% limits.*

4.2. Nelson's Confidence Limits

```
/*   ID  Age  Event              */
/*   Event=-0 : censoring age    */
/*   Event= 1 : recurrence age   */
data bladder;
label group = 'Treatment';
label Age = 'AGE( MONTHS )';
length group $ 10;
input ID$ Age Event;
if ID <= 48 then group = 'Placebo';
else group = 'Thiotepa';
cards;
1     0    -0
2     1    -0
3     4    -0
4     7    -0
5    10    -0
6     6     1
6    10    -0
7    14    -0
8    18    -0
9     5     1
lines of data omitted
82   50    -0
83    4     1
83   24     1
83   47     1
83   50    -0
84   54    -0
85   38     1
85   54    -0
86   59    -0
;
run;

proc reliability data=bladder;
unitid ID;
mcfplot Age*Event(-0) = group / mcfdiff
haxis = 0 to 60 by 10  ;
run;

proc reliability data=bladder;
unitid ID;
mcfplot Age*Event(-0) = group / ncols=1 noconf
vlower=0 vupper=3 nvtick=6
vaxislabel = 'MCF'
haxis = 0 to 60 by 10 ;
run;
```

Figure 4.3(c). *SAS data input and commands.*

extends with constant value to the right of their recurrence up to the next recurrence or censoring age. These limits *do* change at censoring ages, but this is not shown in plots to reduce the clutter. In practice, one may imagine smooth curves through the estimate points and the limits.

- The limits are pointwise limits. That is, they enclose the MCF at a specified single age t with probability $C\%$. In contrast, confidence bands simultaneously enclose the entire MCF over the entire age range of the data with probability $C\%$. Vallarino (1988) presents such bands based on the naive variance of section 4.3.

- The limits allow for observed negative values of events and thus may be negative, too. Such negative limits appear in Figure 4.2.

- The normal approximation for the sampling distribution of $M^*(t)$ is less accurate at both ends of the age range. At the low end, it is usually adequate above the age at which at least 10 events have been accumulated. For early ages with fewer than 10 events, the Poisson limits of section 4.3 are often better.

- The confidence intervals get larger as t increases for two reasons. First, the population standard deviation σ_t of number or cost usually increases as t increases. Second, the number of sample units reaching higher ages is small, resulting in wider intervals.

- These limits require that the input data identify the sample unit associated with each recurrence and censoring age. This is done in the SAS input in Figure 4.3(c), where each event has three data values: patient ID, Age (months in the study), and Value. Value = 1 for a recurrence, and Value = −0 for a censoring age. More generally, Value can be any number, positive or negative.

Software. Except for SPLIDA, the following programs calculate Nelson's confidence limits for $M(t)$ from data with exact ages and right censoring. They provide numerical output and plots. They apply to the number and cost (or value) of recurrences and allow for positive or negative values:

- MCFLIM of Nelson and Doganaksoy (1989). This was used to obtain Figure 4.2.

- The Reliability Procedure in the SAS/QC software of the SAS Institute (1999, pp. 947–951). This was used to obtain Figures 4.3(a) and (b). Numerical confidence limits appear in Table 3.2.

- The JMP software of the SAS Institute (2000, pp. 23–27 and 92–95).

- SPLIDA features developed by Meeker and Escobar (2002) for S-PLUS of Insightful (2001). This uses the variance estimate of Lawless and Nadeau (1995).

- A program developed for General Motors by Robinson (1995) and coded by Ms. Sharon Zielinski.

- The ReliaSoft (2000a, b) Weibull++ software has a recurrence data add-on.

Variance estimates. The limits (4.1) use Nelson's (1995a) estimate $v[M^*(t)]$ for $V[M^*(t)]$, which is unbiased. Robinson (1995) shows that, on rare occasions, the estimate $v[M^*(t)]$ can be negative, especially with a small number of sample units. He proposes another (biased) estimate that is more laborious but always positive. Lawless and Nadeau (1995) offer a simpler (biased) variance estimate that is always positive; it is the same as (4.10) and (4.11) below, but the denominators are replaced, respectively, by i and i'. All these variance estimates yield theoretically correct approximate confidence limits. Except for SPLIDA, the computer programs above use Nelson's variance estimate.

4.3 Naive Confidence Limits

Purpose. This section presents naive approximate confidence limits for the population MCF for the *number* of recurrences from exact age data with right censoring. This section first describes the limitations of these limits. Then it provides formulas for the limits. Finally, it shows how to calculate them step-by-step using the fan motor data of Chapter 1, section 1.2.

Limitations. The naive limits apply only to the *number* of discrete recurrences. They do *not* apply to cost or value data or negative data values. However, these limits for the *number* of recurrences do extend to exact age data with left censoring and gaps and to

4.3. Naive Confidence Limits

interval data (Chapter 5). The naive limits use only the simple tabulation of the data (for example, Table 4.1). In contrast, the correct Nelson limits of section 4.2 require each sample unit's history function. These naive limits require the added, usually dubious, assumption that the population is a nonhomogeneous Poisson process. (This assumption is described in detail in section 4.4 and often is not warranted in applications.) Like the Nelson limits, the naive limits are conditional on the censoring in the sample. The naive intervals are usually shorter than Nelson's and thus make the MCF estimate appear better than it really is. Consequently, one should use the naive limits only as a last resort, for example, when the individual histories are not available. Problems 4.1, 4.2, and 4.3 show how much the correct and naive limits differ.

Notation. The following notation is used in formulas for the naive confidence limits. Suppose that ordered recurrence i is at age t_i, where the true population MCF is $M_i \equiv M(t_i)$ for $i = 1, 2, 3, \ldots$. Also, suppose that for recurrence i at age t_i, r_i is the number at risk. Then the observed increment $m_i = 1/r_i$ of the sample MCF estimates the population increment $(M_i - M_{i-1})$. Here $M_0 \equiv M(0) = 0$. For a nonhomogeneous Poisson process, the true variance of m_i is $V(m_i) = (M_i - M_{i-1})/r_i$, which is estimated with the variance contribution $c_i = v(m_i) = m_i/r_i = 1/r_i^2$.

Variance. The estimate of $M_i \equiv M(t_i)$ is $M_i^* = m_1 + m_2 + \cdots + m_i$. Because a nonhomogeneous Poisson process has independent increments, the increments $m_1, m_2, \ldots, m_i$ are statistically independent. For recurrence i, the true naive variance $V[M_i^*]$ of the MCF estimate M_i^* is

$$V[M_i^*] = V(m_1) + V(m_2) + \cdots + V(m_i)$$
$$= \left[\frac{M_1}{r_1}\right] + \left[\frac{M_2 - M_1}{r_2}\right] + \cdots + \left[\frac{M_i - M_{i-1}}{r_i}\right]. \quad (4.2)$$

Replace the increments $M_1, (M_2 - M_1), \ldots, (M_i - M_{i-1})$ by their estimates $m_1, m_2, \ldots, m_i$ to get the estimate of $V[M^*(t_i)]$:

$$v_i = v[M^*(t_i)] = \left[\frac{m_1}{r_1}\right] + \left[\frac{m_2}{r_2}\right] + \cdots + \left[\frac{m_i}{r_i}\right] = \left[\frac{1}{r_1^2}\right] + \left[\frac{1}{r_2^2}\right] + \cdots + \left[\frac{1}{r_i^2}\right]. \quad (4.3)$$

Limits. The corresponding approximate two-sided naive limits for $M(t_i)$ with $C\%$ confidence are

$$M_i^* - K_C\{v[M_i^*]\}^{1/2} \quad \text{and} \quad M_i^* + K_C\{v[M_i^*]\}^{1/2}, \quad (4.4)$$

where K_C is the $\frac{1}{2}(100 + C)$th standard normal percentile. This naive interval is usually shorter than a correct interval. Thus the naive interval makes the MCF estimate appear more accurate than it actually is. Of course, a poor confidence interval is better than no confidence interval, provided it is viewed skeptically. In industrial applications, these naive limits are often close enough to Nelson's limits when product populations are homogeneous with respect to recurrence rates of individual units. These naive limits for the tumor data are significantly narrower than the correct limits, because the patient population is heterogeneous with respect to recurrence rates of individuals. (In biomedical work, such heterogeneity is modeled with frailty models.) These naive limits do not extend to cost or value data.

Data. The step-by-step calculation of the naive limits is illustrated with data on the *number* of fan motor repairs in heat pumps (section 1.2). These data lack the history of each heat pump. Instead, we have Table 4.1, and we do not know which heat pump had which repair.

Table 4.1. *Calculation of naive confidence limits for fan motors.*

1. Age (days)	2. No. r_i at risk	3. $m_i = 1/r_i$	4. $M_i^* = M_{i-1}^* + m_i$	5. $c_i = m_i/r_i = 1/r_i^2$	6. $v(M_i^*) = v_i = v_{i-1} + c_i$	7. Lower limit	8. Upper limit
141	119	0.008	0.008	7.062E–05	7.062E–05	–0.008	0.025
252+	118						
288+	117						
358+	116						
365+	115						
376+	114						
376+	113						
381+	112						
444+	111						
651+	110						
699+	109						
820+	108						
831+	107						
843	107	0.009	0.018	8.734E–05	1.580E–04	–0.007	0.042
880+	106						
966+	105						
973+	104						
1057+	103						
1170+	102						
1200+	101						
1232+	100						
1269	100	0.010	0.028	1.000E–04	2.580E–04	–0.004	0.059
1355+	99						
1381	99	0.010	0.038	1.020E–04	3.600E–04	0.001	0.075
1471	99	0.010	0.048	1.020E–04	4.620E–04	0.006	0.090
1567	99	0.010	0.058	1.020E–04	5.641E–04	0.012	0.105
1642	99	0.010	0.068	1.020E–04	6.661E–04	0.018	0.119
1646	99	0.010	0.078	1.020E–04	7.681E–04	0.024	0.133
1762+	98						
1869+	97						
1881+	96						
1908	96	0.010	0.089	1.085E–04	8.766E–04	0.031	0.147
1920+	95						
2110+	94						
2261	94	0.011	0.099	1.132E–04	9.898E–04	0.038	0.161
2273	94	0.011	0.110	1.132E–04	1.103E–03	0.045	0.175
2363	94	0.011	0.121	1.132E–04	1.216E–03	0.052	0.189
2419+	93						
2440	93	0.011	0.131	1.156E–04	1.332E–03	0.060	0.203
2562+	92						
2593	92	0.011	0.142	1.181E–04	1.450E–03	0.068	0.217
2615+	91						
2674	91	0.011	0.153	1.208E–04	1.571E–03	0.076	0.231
2710	91	0.011	0.164	1.208E–04	1.691E–03	0.084	0.245
2794+	90						
2815+	89						
2838	89	0.011	0.175	1.262E–04	1.818E–03	0.092	0.259
2946	89	0.011	0.187	1.262E–04	1.944E–03	0.100	0.273
2951	89	0.011	0.198	1.262E–04	2.070E–03	0.109	0.287
2986+	88						
3017+	87						
3087+	86						
3216+	85						
3291+	84						
3296	84	0.012	0.210	1.417E–04	2.212E–03	0.118	0.302
3307	84	0.012	0.222	1.417E–04	2.354E–03	0.127	0.317
3368	84	0.012	0.234	1.417E–04	2.495E–03	0.136	0.332
3391	84	0.012	0.246	1.417E–04	2.637E–03	0.145	0.346
3406+	83						
3440	83	0.012	0.258	1.452E–04	2.782E–03	0.154	0.361
3489	83	0.012	0.270	1.452E–04	2.927E–03	0.164	0.376
3490+	82						
3621+	81						
3631+	80						
3631+	79						
3631+	78						
3631+	77						
3631+	76						
3635+	76	0.013	0.283	1.731E–04	3.101E–03	0.174	0.392
3639+	75						
3648+	74						
3652+	73						

The 73 remaining units are all censored slightly above 3652 days.

4.3. Naive Confidence Limits

Steps. One can systematically calculate the naive confidence limits with the following steps. As in Table 3.3, do the MCF calculation, which appears in columns 1–4 of Table 4.1. Then do the following:

(a) *Variance contribution*: Going down column 5 of Table 4.1, successively calculate the variance contribution of increment m_i as $c_i = m_i/r_i$ for each recurrence i. For example, for the recurrence at 843 days, $c_2 = 0.0093/107 = 8.734\text{E}{-}05$.

(b) *Variance*: Set $v_0 \equiv v[M_0^*] = 0$. Going down column 6, successively calculate the variance estimates of the MCF estimate M_i^* as $v_i \equiv v[M_i^*] = v[M_{i-1}^*] + v(m_i) = v_{i-1} + c_i$ for each recurrence i. For example, for the recurrence at 843 days, $v_2 = v[M_2^*] = 7.062\text{E}{-}05 + 8.734\text{E}{-}05 = 1.580\text{E}{-}04$.

(c) *Limits*: Proceeding down columns 7 and 8, for each recurrence i, calculate the two-sided $C\%$ confidence limits as

$$M_{iL} = M_i^* - K_C\{v[M_i^*]\}^{1/2} = M_i^* - K_C\{v_i\}^{1/2},$$
$$M_{iU} = M_i^* + K_C\{v[M_i^*]\}^{1/2} = M_i^* + K_C\{v_i\}^{1/2}, \quad (4.5)$$

where K_C is the $\frac{1}{2}(100 + C)$th standard normal percentile. For example, for the recurrence at 843 days,

$$M_{2L} = 0.018 - 1.960\{1.580\text{E}{-}04\}^{1/2} = -0.007,$$
$$M_{2U} = 0.018 + 1.960\{1.580\text{E}{-}04\}^{1/2} = 0.042.$$

Such nonparametric lower limits can be negative, which is physically impossible in the fan motor application. Poisson limits, given below, are always positive.

(d) *Plot*: For each recurrence i, plot each estimate M_i^* and its lower and upper limits on the MCF plot above age t_i. Figure 4.4 is this MCF plot with naive limits for the fan motor data. As before, the limits and estimate are staircase functions. In particular, a pair of limits at age t_i extends to their right up to the next recurrence age t_{i+1}. The last pair of limits extends to the longest observed age in the sample, which is slightly above 3652 days for the fan motor data. In practice, one may imagine smooth curves through the estimate and limits.

Software. No commercial software calculates these naive limits. However, they are easy to calculate and plot with a spreadsheet. Table 4.1 and Figure 4.4 were generated by Excel.

Poisson limits. The Nelson and naive approximate limits are suitable when $M^*(t)$ has a sampling distribution "close" to normal. For early ages where there are few recurrences, this distribution is not close to normal. Then it is often better to use a Poisson approximation to the sampling distribution and corresponding limits for the population $M(t)$. For early recurrence i, Poisson two-sided approximate $C\%$ confidence limits for $M_i \equiv M(t_i)$ are

$$M_{il} = \frac{0.5\chi^2\left[\frac{100-C}{2}; 2i\right]}{r_i}, \qquad M_{iu} = \frac{0.5\chi^2\left[\frac{100+C}{2}; 2i+2\right]}{r_i}. \quad (4.6)$$

Here $\chi^2[P; d]$ is the Pth percentile of the chi-square distribution with d degrees of freedom. For example, for recurrence $i = 2$ (at 843 days) of the fan motor data above, these 95% confidence limits are

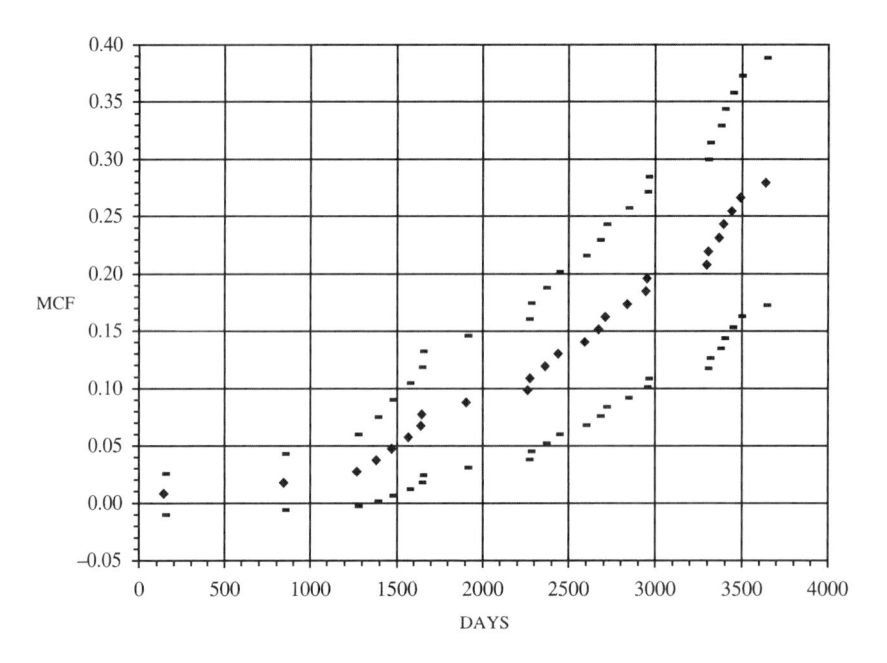

Figure 4.4. *Fan motor MCF and naive 95% confidence limits.*

$$M_{2l} = \frac{0.5\chi^2\left[\frac{100-95}{2}; 2\times 2\right]}{107} = \frac{0.5\chi^2[2.5\%; 4]}{107} = \frac{0.5 \times 0.4844}{107} = 0.002,$$

$$M_{2u} = \frac{0.5\chi^2\left[\frac{100+95}{2}; 2\times 2 + 2\right]}{107} = \frac{0.5\chi^2[97.5\%; 6]}{107} = \frac{0.5 \times 14.45}{107} = 0.067.$$

The corresponding naive limits are −0.007 and 0.042. Poisson limits are usually suitable for $i \leq 10$ and $r_i \geq 20$. Such limits are a staircase function as before. These limits are usually conservative; that is, the probability that they enclose $M(t_i)$ is at least $C\%$.

4.4 Assumptions and Theory

Purpose. This section first reviews the MCF estimate and its assumptions and theory, which are needed background for the confidence limits. Then the section presents the assumptions and theory of Nelson's (1995a) variance of the MCF estimate and confidence limits. In practice, one must assess these assumptions to determine how reliable the confidence limits are. Assumptions are numbered ⟨#⟩ for visibility. Results are numbered {#}.

4.4.1 The MCF Estimate $M^*(t)$

Purpose. This subsection reviews the nonparametric estimate $M^*(t)$ for the population MCF $M(t)$ for *cost* or *number* of recurrences from exact recurrence and censoring ages. As shown in Chapter 3, {3} $M^*(t)$ is an unbiased estimator for $M(t)$.

Notation. The following notation and view of the data are needed to present $M^*(t)$ and derive its variance $V[M^*(t)]$. Figure 4.5 depicts the cost histories of a sample of N units. Each horizontal line depicts a unit cost history, and each × denotes a recurrence. Each vertical dashed line locates the censoring age of a unit. The N units are depicted

4.4. Assumptions and Theory

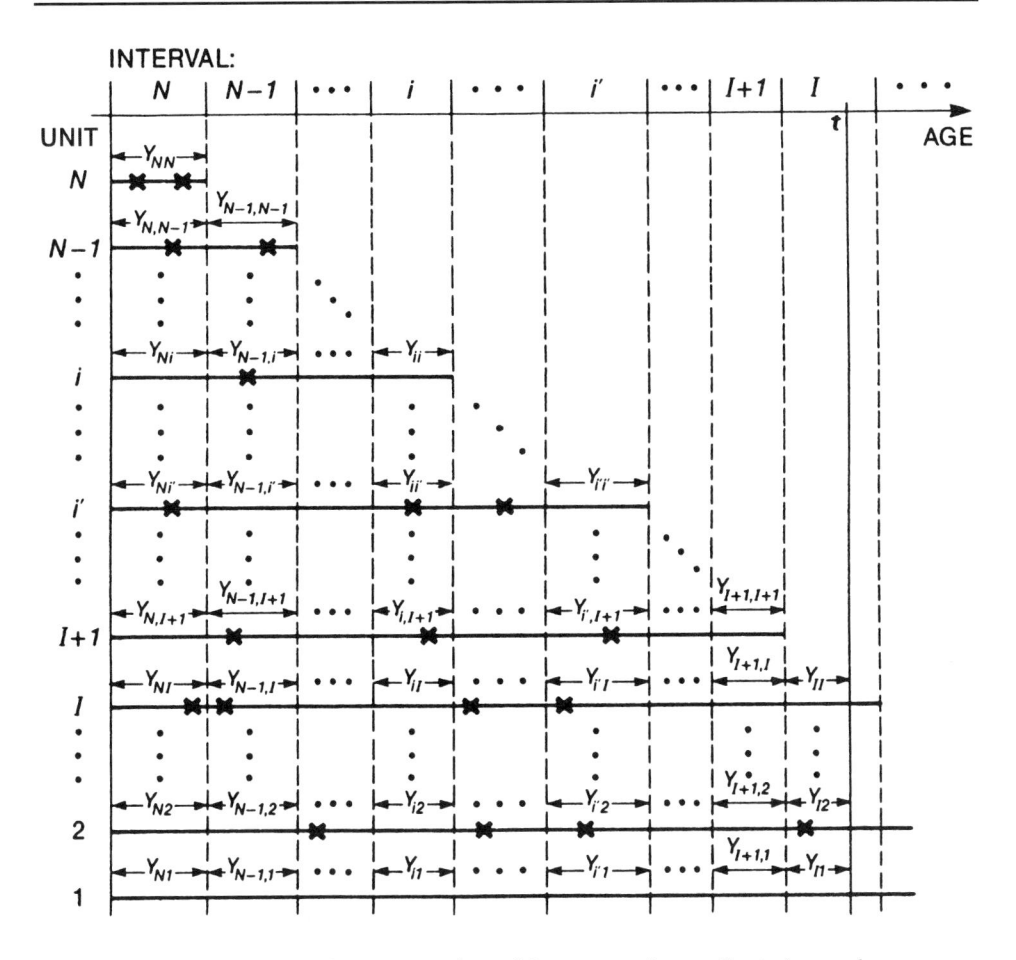

Figure 4.5. *Cost histories with total incremental costs Y_{in} in intervals.*

with the shortest censoring age first, the second shortest censoring age second, etc., and the longest censoring age last. The unit numbers n are backwards: $N, N-1, \ldots, 2, 1$. The N censoring ages divide the observed age range into N intervals. The interval numbers i also are backwards; the earliest interval number is N, the next is $N-1$, etc., and the last is 1. Below we calculate the theoretical variance $V[M^*(t)]$, where t is in interval I and is shown by a solid vertical line. t may be any age in interval I. In practice, we estimate $V[M^*(t)]$ only at recurrence ages, although the estimate changes at censoring ages. Denote the *total incremental recurrence cost* accumulated over all recurrences in interval i for unit n by Y_{in}, $n = 1, 2, \ldots, N$ and $i = 1, 2, \ldots, I-1$. For interval $i = I$, Y_{In} is the total incremental cost accumulated on unit n in interval I up through age t. The arrows $\leftarrow Y_{in} \rightarrow$ in Figure 4.5 span what is accumulated.

$M^*(t)$ formula. The following formula for $M^*(t)$ is used to derive its true variance $V[M^*(t)]$. Here {4} $V[M^*(t)]$ is conditional on the given censoring ages. Examination of the step-by-step calculation of the estimate $M^*(t)$ shows that it is equivalent to the following weighted sum of the Y_{in}:

$$M^*(t) = \begin{array}{cccccccc} \text{unit:} & N & N-1 & N-2 & \ldots & I & \ldots & 1 & \text{interval:} \\ \frac{1}{N} & [Y_{NN} & + Y_{N,N-1} & + Y_{N,N-2} & + \cdots & + Y_{NI} & + \cdots & + Y_{N1}] & N \\ \frac{1}{N-1} & & [Y_{N-1,N-1} & + Y_{N-1,N-2} & + \cdots & + Y_{N-1,I} & + \cdots & + Y_{N-1,1}] & N-1 \\ \frac{1}{N-2} & & & [Y_{N-2,N-2} & + \cdots & + Y_{N-2,I} & + \cdots & + Y_{N-2,1}] & N-2 \\ \vdots & & \ddots & & & \vdots & & \vdots & \vdots \\ \frac{1}{I+1} & & & & & [Y_{I+1,I+1} & + Y_{I+1,I} + \cdots & + Y_{I+1,1}] & I-1 \\ \frac{1}{I} & & & & & [Y_{I,I} & + \cdots & + Y_{I1}] & I. \end{array}$$

(4.7)

The sum in the first row is the total incremental cost of all N units that passed through interval $i = N$; this sum divided by N is the observed average incremental cost per unit in that interval. The sum in the second row is the total incremental cost of all $N - 1$ units that passed through interval $i = N - 1$; this sum divided by $N - 1$ is the observed average incremental cost per unit in that interval. Continuing in the same way, the sum in the last row is the total incremental cost up to age t of all I units that passed through interval $i = I$; this sum divided by I is the observed average incremental cost per unit in that interval up to age t.

$M^*(t)$ assumptions. The assumptions for the MCF estimate $M^*(t)$ from exact age data with right censoring appear in Chapter 3. They are briefly summarized here for convenience:

⟨1⟩ The target population is clearly specified.

⟨2⟩ Simple random sampling of the target population yields the sample.

⟨3⟩ Right censoring of sample histories is random. Equivalently, histories are statistically independent of their censoring ages.

⟨4⟩ Each population history function extends through the age range of the sample data.

⟨5⟩ The population mean $M(t)$ is finite over the range of the data.

⟨6⟩ All recurrence ages are distinct from each other and from the censoring ages.

4.4.2 Variance of the Estimate and Confidence Limits

Overview. This section presents the theoretical variance $V[M^*(t)]$ of the MCF estimate and its unbiased estimate $v[M^*(t)]$. This variance estimate yields approximate confidence limits (4.1) for the MCF of an infinite population. This section first motivates the confidence limits, then states the assumptions, and finally derives the true variance and its estimate used in the confidence limits.

Limits assumptions. For censored history functions, the confidence limits (4.1) require the previous assumptions ⟨1⟩–⟨6⟩ for the estimate $M^*(t)$. The limits also require the following assumptions:

⟨7⟩ The sampling distribution of $M^*(t)$ is close to normal.

4.4. Assumptions and Theory 71

⟨8⟩ All population variances and covariances in the true variance formula (4.9) exist and are finite.

⟨9⟩ The population has an infinite number of units (or histories).

These assumptions are discussed below.

⟨7⟩ **Normality.** The sampling distribution of $M^*(t)$ is assumed to be close to normal. That is, its true percentiles are close to the normal ones in the limits (4.1). This is so for various parametric models for recurrence processes. For example, suppose the observed cumulative number $Y(t)$ of recurrences by time t has a Poisson distribution with true MCF $M(t) = \lambda t$. For λt large (say, greater than 5 or 10), the distribution of the estimate $M^*(t) = Y(t)$ is close to normal. For most counting processes, the normal approximation is crude for ages near 0 where the expected cumulative number of occurrences is fewer than, say, 5 or 10. The normal approximation may be better for costs than for counts for the same process. In practice, the normal confidence limits after a few occurrences are likely adequate when the lower limit is positive. Of course, the accuracy of the normal approximation depends on the population of histories, the age t of interest, and the confidence level $C\%$. Experience suggests that if the population distribution of the cumulative cost or number of recurrences at age t is "close" to normal, then so is the sampling distribution of $M^*(t)$. Also, many population distributions are closer to normal at higher ages; the mixture population in Figure 4.7 later is an exception. Also, the approximation is poorer for higher confidence levels $C\%$.

⟨9⟩ **Infinite population.** It is assumed that the population consists of an infinite number of units (histories). This standard assumption allows us to use the usual properties of variances and covariances in formulas below. In practice, the population can usually be regarded as infinite if the sample size is less than 10% of the population. Robinson (1995) extended the variance (4.9) to finite populations, where the sample can be the entire finite population.

True variance. In evaluating the variance of (4.7), the censoring ages are regarded as given, and {4} the true $V[M^*(t)]$ is conditional on those ages. The variance $V[M^*(t)]$ of (4.7) is the usual variance of a sum. Thus $V[M^*(t)]$ contains the population variances $V(Y_{in})$ for interval i of all population Y_{in} values ($n = 1, 2, \ldots, N$) and the population covariances $V(Y_{in}, Y_{i'n})$ for intervals i and i' of all population pairs of observations $(Y_{in}, Y_{i'n})$, $n = 1, 2, \ldots, N$. Such a covariance reflects the *population* autocorrelation between *incremental* costs in intervals i and i', not the autocorrelation of a single unit, which has no meaning under the population model here. For sample units n and n', all covariances $V(Y_{in}, Y_{i'n'}) = 0$ because sample units n and n' are statistically independent; this is a result of assuming ⟨2⟩ simple random sampling from ⟨9⟩ an infinite population. Then the variance of $M^*(t)$ is

$$
\begin{aligned}
M^*(t) = \quad & \frac{1}{N^2} && [V(Y_{NN}) + V(Y_{N,N-1}) + V(Y_{N,N-2}) && + \cdots && + V(Y_{NI}) && + \cdots + V(Y_{N1})] \\
& \frac{1}{(N-1)^2} && [V(Y_{N-1,N-1}) + V(Y_{N-1,N-2}) && + \cdots && + V(Y_{N-1,I}) && + \cdots + V(Y_{N-1,1})] \\
& \frac{1}{(N-2)^2} && \quad\quad\quad [V(Y_{N-2,N-2}) && + \cdots && + V(Y_{N-2,I}) && + \cdots + V(Y_{N-2,1})] \\
& \vdots && \quad\quad\quad\quad\quad \ddots && && \vdots && \quad\quad \vdots \\
& \frac{1}{(I+1)^2} && && && [V(Y_{I+1,I+1}) + V(Y_{I+1,I}) && + \cdots + V(Y_{I+1,1})] \\
& \frac{1}{I^2} && && && [V(Y_{I,I}) && + \cdots + V(Y_{I1})]
\end{aligned}
$$

$$+ \frac{2}{N(N-1)} \sum_{n=1}^{N-1} V(Y_{Nn}, Y_{N-1,n})$$

$$+ \frac{2}{N(N-2)} \sum_{n=1}^{N-2} V(Y_{Nn}, Y_{N-2,n})$$

$$+ \frac{2}{N(N-3)} \sum_{n=1}^{N-3} V(Y_{Nn}, Y_{N-3,n})$$

$$\vdots \qquad \vdots$$

$$+ \frac{2}{N(I+1)} \sum_{n=1}^{I+1} V(Y_{Nn}, Y_{I+1,n})$$

$$+ \frac{2}{NI} \sum_{n=1}^{I} V(Y_{Nn}, Y_{I,n})$$

$$\dots\dots\dots\dots\dots\dots\dots\dots\dots\dots\dots\dots\dots\dots\dots\dots\dots\dots\dots\dots\dots\dots \qquad (4.8)$$

$$+ \frac{2}{(N-1)(N-2)} \sum_{n=1}^{N-2} V(Y_{N-1,n}, Y_{N-2,n})$$

$$+ \frac{2}{(N-1)(N-3)} \sum_{n=1}^{N-3} V(Y_{N-1,n}, Y_{N-3,n})$$

$$+ \frac{2}{(N-1)(N-4)} \sum_{n=1}^{N-4} V(Y_{N-1,n}, Y_{N-4,n})$$

$$\vdots \qquad \vdots$$

$$+ \frac{2}{(N-1)(I+1)} \sum_{n=1}^{I+1} V(Y_{N-1,n}, Y_{I+1,n})$$

$$+ \frac{2}{(N-1)I} \sum_{n=1}^{I} V(Y_{N-1,n}, Y_{I,n})$$

$$\dots\dots\dots\dots\dots\dots\dots\dots\dots\dots\dots\dots\dots\dots\dots\dots\dots\dots\dots\dots\dots\dots$$

$$\vdots \qquad \vdots$$

$$\dots\dots\dots\dots\dots\dots\dots\dots\dots\dots\dots\dots\dots\dots\dots\dots\dots\dots\dots\dots\dots\dots$$

$$+ \frac{2}{(I+1)I} \sum_{n=1}^{I} V(Y_{I+1,n}, Y_{In}).$$

The first block of terms consists of the individual variances of each of the Y_{in} terms of (4.7). The second block consists of the covariances between incremental costs in the interval $i = N$ and those in each of the subsequent intervals $i = N-1, i = N-2, \ldots, i = I$. The third block consists of the covariances between incremental costs in interval $i = N-1$ and those in each of the subsequent intervals $i = N-2, \ldots, i = I$, and so on through the last block, which consists of the covariances between incremental costs in interval $i = I+1$ and those

4.4. Assumptions and Theory

in the subsequent interval $i = I$ up to age t.

Simplify. Equation (4.8) simplifies as follows. For interval $i = N$, the Y_{NN}, $Y_{N,N-1}$, ..., Y_{N1} (appearing in the first row of the first block of (4.8)) are N independent observations from the *same* incremental cost distribution of the population for that interval. Thus $V(Y_{NN}) = V(Y_{N,N-1}) = \cdots = V(Y_{N1})$. Denote the common variance by $V(Y_{Nn})$. Then the sum of the first row is $N \cdot V(Y_{Nn})$. Similarly, in the second row (for interval $i = N - 1$) of the first block, the $N - 1$ variances have a common value $V(Y_{N-1,n})$ and a sum of $(N-1)V(Y_{N-1,n})$, and so on; in the last row (for interval $i = I$) of the first block, the I variances have a common value $V(Y_{In})$ and a sum of $I \cdot V(Y_{In})$. Subsequent covariance terms can be combined similarly. In particular, the covariance terms in a sum in a single row of (4.8) are all equal. For example, the first such row has $N - 1$ covariances with a common value $V(Y_{Nn}, Y_{N-1,n})$ and a sum of $(N-1)V(Y_{Nn}, Y_{N-1,n})$. After all such substitutions, formula (4.8) simplifies to {5} the desired theoretical variance

$$V[M^*(t)] = \frac{1}{N}V(Y_{Nn}) + \frac{1}{N-1}V(Y_{N-1,n}) + \frac{1}{N-2}V(Y_{N-2,n}) + \cdots + \frac{1}{I}V(Y_{In})$$
$$+ \frac{2}{N}[V(Y_{Nn}, Y_{N-1,n}) + V(Y_{Nn}, Y_{N-2,n}) + \cdots + V(Y_{Nn}, Y_{In})]$$
$$+ \frac{2}{N-1}[V(Y_{N-1,n}, Y_{N-2,n}) + \cdots + V(Y_{N-1,n}, Y_{In})]$$
$$+ \cdots$$
$$+ \frac{2}{I+1}[V(Y_{I+1,n}, Y_{In})].$$
(4.9)

Covariance. Figure 4.6 is a crossplot of a bivariate population of $(Y_{in}, Y_{i'n})$ values, which are the total incremental costs in intervals i and i' for all population units. A point's horizontal coordinate Y_{in} is the unit's total incremental cost in interval i (column i of Figure 4.5), and the point's vertical coordinate is the unit's total incremental cost in interval i' (column i' of Figure 4.5). The population covariance $V(Y_{in}, Y_{i'n})$ in (4.9) is just the covariance of the population points in Figure 4.6. For most populations, many of the incremental costs are zero. That is, many points appear on the axes and at the origin, as in Figure 4.6.

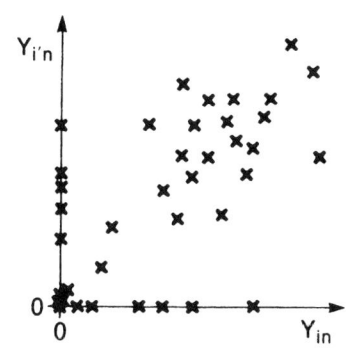

Figure 4.6. *Crossplot of a bivariate population of incremental costs for intervals i and i'.*

Variance estimate. Each true variance $V(Y_{in})$ of (4.9) is estimated from the Y_{in} data in Figure 4.5 as follows. The i sample incremental costs $Y_{ii}, Y_{i,i-1}, \ldots, Y_{i1}$, observed in interval i, appear in column i of Figure 4.5. These observed costs are a random sample from that incremental cost distribution. Thus their sample variance is an unbiased estimate of the population variance $V(Y_{in})$; namely,

$$v(Y_{in}) = \sum_{n=1}^{i} \frac{(Y_{in} - \bar{Y}_{i\cdot})^2}{i-1}, \qquad (4.10)$$

where $\bar{Y}_{i\cdot} = (Y_{ii} + Y_{i,i-1} + \cdots + Y_{i1})/i$. Each true population covariance $V(Y_{in}, Y_{i'n})$, $i > i'$, of (4.9) is estimated from the Y_{in} data in Figure 4.5 as follows. The i' pairs of sample incremental costs $(Y_{ii'}, Y_{i'i'})$, $(Y_{i,i'-1}, Y_{i',i'-1})$, ..., $(Y_{i1}, Y_{i'1})$ are observed on the i' units in intervals i and i'. These pairs appear in columns i and i' of Figure 4.5. These i' pairs are a random sample from the bivariate population of incremental costs in Figure 4.6. Thus the sample covariance of the i' observed pairs is an unbiased estimate for the population covariance $V(Y_{in}, Y_{i'n})$; namely,

$$v(Y_{in}, Y_{i'n}) = \sum_{n=1}^{i'} \frac{(Y_{in} - \bar{Y}_{i\cdot})(Y_{i'n} - \bar{Y}_{i'\cdot})}{i' - 1}, \quad i' < i. \qquad (4.11)$$

The unbiased estimates (4.10) and (4.11) substituted into (4.9) yield {6} an unbiased estimate $v[M^*(t)]$ for the true $V[M^*(t)]$. Then $v[M^*(t)]$ used in (4.7) yields two-sided normal approximate $C\%$ confidence limits for the population $M(t)$. Robinson (1995) extended (4.9) and these limits to finite populations. He also showed that the unbiased estimate $v[M^*(t)]$ can be negative with an artificial data set and a few simulated data sets out of thousands. He investigated a positive (biased) variance estimate; he and the author have not seen negative estimates with real data sets. Lawless and Nadeau (1995) provide another positive (biased) estimate of the variance (4.9).

Independence. In the derivation above, Y_{in} and $Y_{i'n'}$ are *statistically* independent for $n \neq n'$, that is, for different sample units. This is so because ⟨2⟩ simple random sampling from ⟨9⟩ an infinite population is used to obtain the sample units, and ⟨3⟩ the given censoring ages are randomly assigned to them. This is so even if some sample units may have physically correlated cost histories. Such physical correlation may occur among units in a common operating environment or from a certain segment of production. In practice, many samples are not simple random ones.

Not assumed. Much modeling of recurrence data involves parametric stochastic point processes. Such models yield estimates and confidence limits under more restrictive and often dubious assumptions. Assumptions used by other authors to model and to simplify analysis of such recurrence data appear in section 3.6. Such assumptions are not used here.

Independent increments. For approximate nonparametric confidence limits for an MCF, Rigdon and Basu (2000), Ascher and Feingold (1984, pp. 21 and 30), Engelhardt (1995), Vallarino (1988), and other authors assume that ⟨10⟩ the $Y_n(t)$ are from a nonhomogeneous Poisson process. They use this strong, simplifying (and often unrealistic) assumption to make their mathematics tractable. Such a counting process has independent increments; this implies that $V(Y_{in}, Y_{i'n}) = 0$ for all i and i' in (4.9). This assumption is needed for the naive limits. The author knows of no hypothesis test for independent increments. It would be useful to test the weaker assumption that all covariances $V(Y_{in}, Y_{i'n})$ equal 0. It may be possible to develop such a test analogous to tests from multivariate theory for a diagonal covariance matrix. Then the null hypothesis model would be a nonhomogeneous Poisson process.

Dependent increments. Figure 4.7 depicts a common situation with dependent increments. Suppose that the (infinite) population has a mixture of history functions (cumulative

4.4. Assumptions and Theory

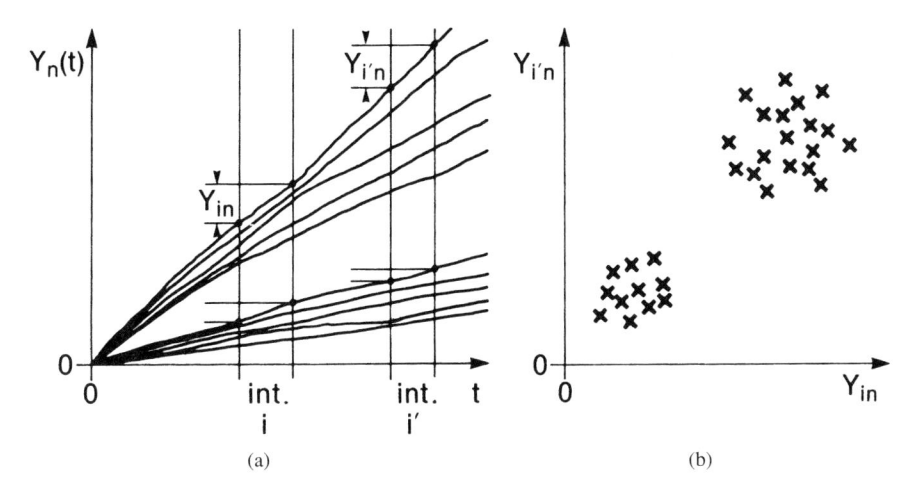

Figure 4.7. (a) *History functions of a mixture of two Poisson processes.* (b) *Crossplot of a correlated bivariate population of costs* $(Y_{in}, Y_{i'n})$.

number of recurrences) generated by two Poisson processes (each with independent increments). One process has a high recurrence rate and the other has a low rate, say, corresponding to severe and mild environments. Figure 4.7(a) shows the history functions as curves for easy viewing, rather than as step functions. Consider the pair of the increments $(Y_{in}, Y_{i'n})$ of the function for population unit n in intervals i and i'. All such population pairs are crossplotted in Figure 4.7(b). The figure shows that the population covariance $V(Y_{in}, Y_{i'n}) \neq 0$. This mixture population does *not* have independent increments, even though each process does. Confidence limits of other authors do not apply here, whereas those based on (4.9) require minimal assumptions and do apply. In practice, many populations are mixtures. The patients with bladder tumors are such a population. Frailty models are used to model such mixtures.

Extensions. The Nelson confidence limits for exact age data readily extend as follows:

- MCFs of two independent samples can be compared pointwise at any age t with a confidence interval, as described in Chapter 7.

- Censoring may be more complex than the right censoring above. Data may contain left and right censoring and gaps. The data in Table 1.4 on the replacement of compressors in air conditioners have initial gaps (left censoring); some were put on service contract after a year or two in service, and prior repair records were not available. Also, for example, patients may enter and leave a medical study of a disease any number of times and thus have gaps in their histories.

- The history functions above are regarded as step functions generated by discrete events at exact ages. The theory extends to continuous history functions, for example, for cumulative uptime of repairable units. In practice, one usually would calculate such MCF estimates and confidence limits at equally spaced points over the age range.

Resampling limits. The Nelson confidence limits and others mentioned above require that the sampling distribution of $M^*(t)$ be approximately normal. The jackknife, the bootstrap, and other resampling methods could be employed to obtain an empirical sampling distribution for the limits and do not entail a normal approximation. For count data, such limits are always positive, whereas the normal ones may have a negative lower limit. Allison (1996) studies such limits, and Cohen et al. (1998) obtain such limits for arrest data.

Wang, Qin, and Chiang (2001) obtain such limits for patient data with informative censoring. General references include the following:

- M. R. CHERNICK (1999), *Bootstrap Methods: A Practitioner's Guide*, John Wiley, New York (contains 1600 references).

- A. C. DAVISON AND D. V. HINKLEY (1997), *Bootstrap Methods and Their Application*, Cambridge University Press, Cambridge, UK.

- B. EFRON AND R. J. TIBSHIRANI (1994), *An Introduction to the Bootstrap*, CRC Press, Boca Raton, FL.

Stratify. In some applications, it may be possible to reduce the variance $V[M^*(t)]$ of the population MCF estimate. This can be done by stratifying the population into homogeneous subpopulations whose MCF estimates have smaller variances. Then we combine the stratum MCF estimates to obtain an estimate for the population MCF, whose variance is smaller than $V[M^*(t)]$. Cochran (1977) discusses stratification.

Problems

4.1. Transmission data. Use the automatic transmission data from Table 3.1.

(a) Calculate the naive 95% confidence limits and plot them on Figure 3.1.

(b) Calculate the Poisson 95% confidence limits and plot them on Figure 3.1.

(c) Compare the three sets of limits and state which you prefer at each recurrence and why.

Use the manual transmission data from Problem 3.1.

(d) Calculate the naive 95% confidence limits and plot them on Figure 4.2.

(e) Calculate the Poisson 95% confidence limits and plot them on Figure 4.2.

(f) Compare the three sets of limits and state which you prefer at each recurrence and why.

4.2. Tumor data. Use the Thiotepa or placebo data (Tables 3.6 and 1.2, respectively).

(a) Calculate the naive 95% limits for $M(t)$.

(b) Plot the limits on Figures 4.3(a)–4.3(a).

(c) Describe how the naive limits compare with the Nelson limits in the figure. The data do not have independent increments.

4.3. Proschan data. Use the Proschan data on air conditioner systems from Problems 2.2 and 3.3.

(a) Calculate and plot the MCF estimate and Nelson 95% confidence limits. In view of the limits, is the straight line plausible?

(b) Calculate and plot the naive 95% confidence limits. Note that here the naive limits are shorter than the correct limits, since the population is better modeled as a mixture of Poisson processes, with a different recurrence rate for each system, according to Proschan (2000).

4.4. Fan motors. Use the fan motor data in Table 4.1.

(a) Calculate the Poisson 95% confidence limits for the first five repairs.

(b) Plot these limits on Figure 4.4.

(c) Comment on how the two sets of limits compare.

4.5. Circuit breakers. Use the circuit breaker data in Problem 3.5.

(a) Use a computer program to calculate Nelson's 95% confidence limits for the MCF. Plot the MCF and limits. Interpret the plot.

(b) Calculate the naive 95% limits and plot them on the same plot.

(c) Calculate the Poisson 95% limits and plot them on the same plot.

(d) Compare the three sets of limits and recommend which to use at each recurrence age and why.

4.6. Chronic granulomatous disease. Fleming and Harrington (1991, pp. 162–163 and 376–383) give data on recurrences of serious infections in patients with chronic granulomatous disease (CGD). Patients were randomly assigned to gamma interferon treatment or to placebo treatment.

(a) Calculate and plot the MCF and 95% confidence limits for each treatment group, ignoring covariates.

(b) Compare the two MCFs. Do you think that the difference is convincing? Why?

(c) Use the covariates to divide the data into covariate groups, for example, using or not using corticosteroids; using or not using antibiotics; male or female, etc. Calculate and plot the MCF for each group. Note the differences that you think are convincing and why.

4.7. Your data. Choose a set of recurrence data, preferably your own.

(a) Use a computer program to calculate and plot the sample MCF and confidence limits.

(b) Interpret your plot.

Chapter 5
MCF Estimate and Limits for Interval Age Data

5.1 Introduction

Purpose. This chapter shows how to calculate a nonparametric estimate and naive confidence limits for the population MCF from *interval* age data and how to plot and interpret the estimate and limits. This useful plot provides most of the information sought from such recurrence data. It plays as key a role for recurrence data as do probability plots for life data. The necessary background for this chapter is contained in Chapters 3 and 4.

Interval data. Censoring and recurrence ages are usually grouped into intervals for convenience, for example, into months or years. For example, the automotive industry often uses months. Also, the childbirth data of section 1.4 are an example of such data because birthdates were grouped. Interval data consist of just the number of recurrences and the number of censoring ages in each interval, which compresses the data set. For example, the defrost control data in Problem 5.2 below contains 22,914 controls, and the data are summarized with (1) the number of controls replaced and (2) the number censored in each of 29 months. This reduces the data on 22,914 controls to $2 \times 29 = 58$ numbers. Of course, grouping the recurrence and censoring ages into intervals results in less accurate estimates, but the reduction in accuracy is usually small compared with the resulting convenience.

Chapter overview. The following sections of this chapter deal with estimating the MCF for such interval age data, which are described in Chapter 1, and the practical and theoretical issues:

5.2 *MCF estimate*.

5.3 *Confidence limits*.

Problems.

5.2 MCF Estimate

Purpose. This section shows how to calculate, plot, and interpret the MCF estimate for the number of recurrences from interval age data with right censoring. The calculation is illustrated with the childbirth data discussed in Chapter 1. This section also presents extensions of the MCF estimate and discusses some issues.

Interval data. The childbirth data for men from Table 1.5 illustrate the MCF estimate and appear in Table 5.1. For each interval i (column 1), Table 5.1 shows

- the age range of the interval (column 2),
- the number of recurrences R_i (column 3, children born to men) in that interval, and
- the number C_i of men (column 4) with current (censoring) ages in that interval.

The ages were grouped into convenient intervals (0–20, 20–22, 22–24, ..., 60–99). For the MCF estimate below, the age intervals need not have equal length, but ⟨6'⟩ all sample units must be grouped into a common set of intervals, like the childbirth data. Thall and Lachin (1988) deal with recurrence data where different units have different age intervals. Lawless and Zhan (1998) give an MCF estimate consisting of straight line segments.

Table 5.1. *Sample MCF calculation for men (∗∗ indicates omitted estimate).*

1. Interval i	2. Endpoints $t_{i-1} - t_i$	3. Number of recurrences R_i	4. Number censored C_i	5. Enter $N_i = N_{i-1} - C_{i-1}$	6. Observed increment $m_i = R_i/(N_i - 0.5 \times C_i)$	7. Cumulative $M_i^* = M_{i-1}^* + m_i$
0	0					0.000
1	0–20			59	$0/(59 - 0.5 \times 0) = 0.000$	0.000
2	20–22	3	1	59	$3/(59 - 0.5 \times 1) = 0.051$	0.051
3	22–24	3		58	$3/(58 - 0.5 \times 0) = 0.052$	0.103
4	24–26	3	1	58	$3/(58 - 0.5 \times 1) = 0.052$	0.155
5	26–28	5	1	57	$5/(57 - 0.5 \times 1) = 0.088$	0.243
6	28–30	6	3	56	$6/(56 - 0.5 \times 3) = 0.110$	0.353
7	30–32	11	3	53	$11/(53 - 0.5 \times 3) = 0.214$	0.567
8	32–34	6	5	50	$6/(50 - 0.5 \times 5) = 0.126$	0.693
9	34–36	9	2	45	$9/(45 - 0.5 \times 2) = 0.205$	0.898
10	36–38	6	3	43	$6/(43 - 0.5 \times 3) = 0.145$	1.043
11	38–40	2	3	40	$2/(40 - 0.5 \times 3) = 0.052$	1.095
12	40–45	7	7	37	$7/(37 - 0.5 \times 7) = 0.209$	1.304
13	45–50	2	14	30	$2/(30 - 0.5 \times 14) = 0.087$	1.391
14	50–55		5	16	$0/(16 - 0.5 \times 5) = 0.000$	1.391
15	55–60		8	11	$0/(11 - 0.5 \times 8) = 0.000$	1.391
16	60–99		3	3	∗∗	∗∗
	Total:	63	59	0		

Information sought. The data analyses below answer the following questions:

- How do the childbirth rates of male and female statisticians compare as a function of age? Common experience suggests that women tend to have children at younger ages than men do. Also, experience suggests that men continue having children into older ages since women are limited by menopause.

- How many children on the average do statisticians ultimately have? To sustain a population, this number needs to be about 2.1. Two children are needed to replace the statistician and partner, and the excess 0.1 accounts for children who die before childbearing age.

- Do men and women ultimately have the same average number of children? Of course, common knowledge answers this.

Give thought to what you expect the data to show.

Estimate. Calculate the sample estimate of the MCF from interval data with the following steps. This calculation for men appears in Table 5.1. In practice, this calculation is done with a computer program or a spreadsheet.

5.2. MCF Estimate

(a) *Tabulate data*: Arrange the age data in order as shown in columns 1–4 of Table 5.1. Sum the number of recurrences R_i (births here) in column 3; the total is 63 births. Sum the numbers C_i censored in column 4 to get the number of sample units, 59 men.

(b) *Number entering*: In row 1 of column 5, enter the sample size N_1 (59 here) for age 0. Then go down column 5 successively calculating the number $N_i = N_{i-1} - C_{i-1}$ of units that entered interval i. For example, $N_3 = 59 - 1 = 58$. Blanks in the columns are treated as 0's throughout. Thus $N_4 = 58 - 0 = 58$.

(c) *Increment*: Going down column 6, successively calculate the observed increment $m_i = R_i/(N_i - 0.5 \times C_i)$, which is the average number of recurrences per sample unit over interval i. For example, $m_2 = 3/(59 - 0.5 \times 1) = 0.051$. R_i is the number of recurrences in interval i. The denominator $(N_i - 0.5 \times C_i)$ approximates the observed number at risk in that interval. The C_i censored units are treated as if they went halfway through interval i on the average. Since no units went entirely through the *last* interval 15 (60–99), its denominator $(3 - 0.5 \times 3)$ is biased high, and thus the increment estimate is biased low by an unknown amount. Thus the corresponding estimates in the last interval are replaced with ∗∗ in Table 5.1. Robinson and McDonald (1991) deal with this last interval issue. Not used here, the observed recurrence rate for interval i is $m_i/(t_i - t_{i-1})$.

(d) *MCF*: In column 7, set $M_0^* = 0$. Then go down column 7, successively calculating each estimate $M_i^* = M_{i-1}^* + m_i$ for $M(t_i)$, where t_i is the upper endpoint of interval i. For example, $M_2^* = 0 + 0.051 = 0.051$ at $t_2 = 22$ years, and $M_3^* = 0.051 + 0.052 = 0.103$ at $t_3 = 24$ years.

(e) *Plot*: On a suitable graph grid (usually a square grid), plot each MCF estimate M_i^* versus its *upper* endpoint t_i. This plot of the men's MCF appears in Figure 5.1. For example, the point for interval 12 has the coordinates (50, 1.391).

Figure 5.1 also shows the women's MCF calculated from Table 1.5. Interpretations follow.

Interpretation. In Figure 5.1, the plots for men and women look smooth and reasonable. The sample individuals started having children around age 22, later than the general population. In this small sample, men started having children earlier, contrary to common expectation. The rate of births (derivative) for men rises to a peak in the early thirties. The rate for women is relatively constant from 25 to 40. At older ages, men continue having children and women do not, of course, due to menopause. We usually visualize a smooth curve through the points since most population MCFs are smooth curves. The sample MCF extends to the oldest sample age, beyond 60 years for men and women. Thus both plots ultimately level off near 1.4 children per statistician, equal for men and women, as expected. This is below the 2.1 or so needed to sustain a population. Such a low birth rate is typical of many well-educated groups.

Computer programs. There are no computer programs that calculate the sample MCF for interval age data. In 2002, SAS Institute plans to include such features in the SAS/QC Reliability Procedure. The calculation is simple and easy to do with a spreadsheet. The preceding calculation and plot were generated using Excel.

Constant rate. The estimate m_i of the MCF increment in interval i implicitly assumes that ⟨7′⟩ the recurrence or cost rate is constant over the interval. Thus the MCF estimate M_i applies to the upper endpoint t_i of the interval. In this spirit, one could connect the plotted points (t_i, M_i^*) with straight line segments to estimate the MCF, as in Figure 5.1. The sample MCF is not regarded as a staircase function here. Alternatively, one could draw or imagine a

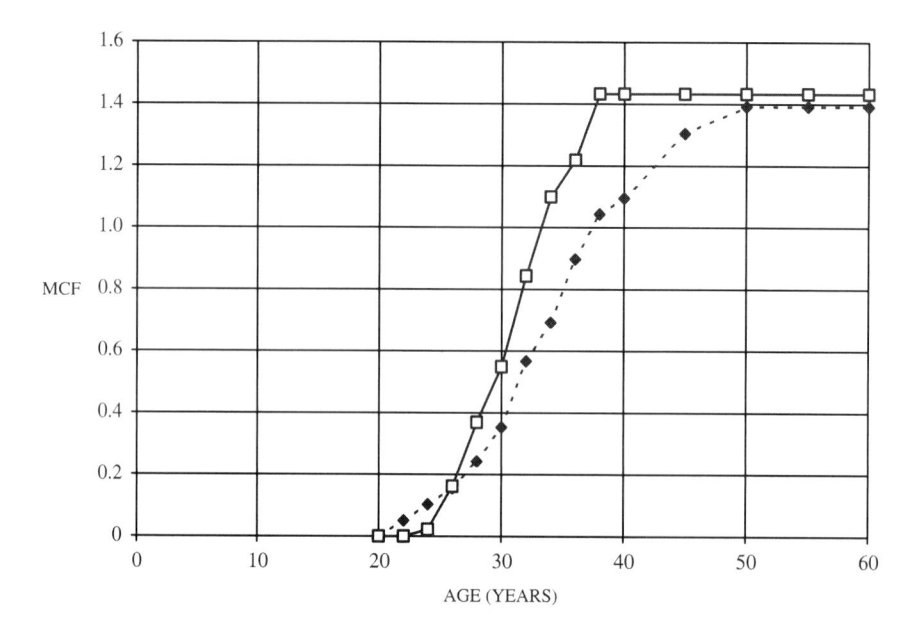

Figure 5.1. *Plot of sample MCFs for childbirth data* (□ = *women*, ♦ = *men*).

smooth curve through the points. Also, one can present only the plotted points. The choice of how to smooth the plot depends on personal taste or the application. The adjustment $-0.5C_i$ for the number of censored units in the denominator of m_i employs assumption $\langle 7' \rangle$. The actuarial estimate of a life distribution from interval data has a similar adjustment, which appears in Nelson (1982, pp. 150–154).

Bias. The MCF estimate for exact age data is unbiased. This estimate for interval data is biased; however, the bias is likely to be small compared to the statistical randomness in most applications.

Data points. In previous MCF plots for exact age data, each plotted point corresponds to a recurrence. For interval age data, each plotted point corresponds to an interval and may represent any number of recurrences. Thus, in interpreting a plot, one needs to be aware that the amount of data is not apparent in the MCF plot for interval data. Some analysts may prefer to plot a point for each recurrence and to spread the points evenly over their interval. The tumor data are interval data with one-month intervals. In Chapter 3, these data were treated as exact age data, and those tumor plots contain a point for each recurrence.

Cost/value data. The MCF estimate above extends to cost or value data in the obvious way. Namely, the observed total cost or value Y_i for all recurrences in interval i takes the place of the number R_i of recurrences in the MCF calculations. Extending this estimate to continuous history functions is more difficult.

Censoring. The MCF estimate above applies to right censored histories. The estimate extends in the obvious way to histories with left censoring and gaps. Also, the estimate applies to data with a mix of events (Chapter 6).

Assumptions. The MCF estimate here for interval age data requires all assumptions $\langle 1 \rangle$–$\langle 5 \rangle$ in Chapter 3 for the corresponding MCF estimate for exact age data. Namely, the assumptions are as follows:

$\langle 1 \rangle$ The target population is clearly specified.

⟨2⟩ The sample was obtained with simple random sampling of the target population.

⟨3⟩ Censoring of the sample histories is random.

⟨4⟩ All sample histories, if not censored, would extend to any observed age of interest.

⟨5⟩ The MCF exists mathematically at any observed age of interest.

In addition, the MCF estimate here for interval data requires further assumptions:

⟨6′⟩ All sample units are grouped into a common set of intervals.

⟨7′⟩ The population recurrence rate is constant over each interval.

⟨8′⟩ The adjustment $-0.5C_i$ in the increment properly adjusts for the unknown exact censoring ages in interval i.

In practice, the effect of departures from these assumptions must be evaluated to assess the adequacy of the MCF estimate here. For example, when no units go entirely through the last interval I, the adjustment ⟨8′⟩ yields a denominator $(N_I - 0.5 \times C_I)$ that is biased high, and M_I^* is underestimated and is best omitted.

5.3 Confidence Limits

Purpose. This section describes how to calculate approximate confidence limits for the MCF from interval age data. This section first presents Nelson's correct limits, which apply to the number or cost of recurrences. The correct limits are complicated and require a computer program. Finally, this section presents naive limits, which apply only to the number of recurrences and often yield intervals that are too narrow. These naive limits are presented first in formulas and then step-by-step in a table as an algorithm.

Correct limits. Correct confidence limits for the MCF like those in Chapter 4 have not been developed for interval age data. Correct limits would require *all of the individual sample histories* instead of the simple tabulation of just the number of recurrences and the number of censoring ages in each interval. One can approximate the Nelson limits of Chapter 4 by treating the interval data as exact age data. To do this, treat each interval of a unit as an event whose value is the total number (or total value) of that unit's recurrences in that interval. Use the interval's upper endpoint as the event's age. These pseudoexact data can be analyzed with a computer program as described in Chapter 4. This was done with the tumor recurrence data. Table 3.2 of Chapter 3 displays the MCF calculation, and Figure 3.2 displays the MCF plot with such confidence limits, which are a little narrower than correct ones. This approximation yields best results if the interval widths are small compared to the age range of the data. Other limits using the interval data and all individual sample histories can be obtained with bootstrapping (section 4.4). Both such limits for interval data apply to cost (or value) data, as well as to the number of recurrences.

Naive limits. The naive confidence limits of Chapter 4 for the *number* of recurrences extend to interval age data as follows. The naive limits use only the simple tabulation of the data (for example, Table 5.1) and do not require all the individual unit history functions. As in Chapter 4, these naive limits employ the assumption that the population is a nonhomogeneous Poisson process, which is discussed in Chapter 8.

Notation. Suppose that interval i has endpoints $(t_{i-1}, t_i]$ and that the true population MCF at age t_i is $M_i \equiv M(t_i)$ for $i = 1, 2, 3, \ldots$. Here $M_0 \equiv M(0)$. Also, suppose that, for interval i, N_i is the observed total number of sample units entering the interval, R_i is the total number of recurrences in the interval, and C_i the number of units censored in the interval. The observed increment $m_i = R_i/(N_i - 0.5C_i)$ estimates the population increment $M_i - M_{i-1}$ over interval i. Its true variance is $V(m_i) \cong (M_i - M_{i-1})/(N_i - 0.5C_i)$, and $v(m_i) = m_i/(N_i - 0.5C_i)$ estimates $V(m_i)$.

Variance. The estimate of $M_i \equiv M(t_i)$ is $M_i^* = m_1 + m_2 + \cdots + m_i$. Because of the assumption of independent increments, $m_1, m_2, \ldots, m_i$ are statistically independent. For interval i, the "true" naive variance $V[M_i^*]$ of the estimate M_i^* is

$$V[M_i^*] = V(m_1) + V(m_2) + \cdots + V(m_i)$$

$$\cong \left[\frac{M_1}{N_1 - 0.5C_1}\right] + \left[\frac{M_2 - M_1}{N_2 - 0.5C_2}\right] + \cdots + \left[\frac{M_i - M_{i-1}}{N_i - 0.5C_i}\right].$$

Replace the increments $M_1, (M_2 - M_1), \ldots, (M_i - M_{i-1})$ by their estimates $m_1, m_2, \ldots, m_i$ to get the estimate of $V[M_i^*]$:

$$v[M_i^*] \cong \left[\frac{m_1}{N_1 - 0.5C_1}\right] + \left[\frac{m_2}{N_2 - 0.5C_2}\right] + \cdots + \left[\frac{m_i}{N_i - 0.5C_i}\right].$$

The square root of this is the estimate of the naive standard error of M_i^*.

Limits. The corresponding approximate two-sided naive limits for $M(t_i)$ with $C\%$ confidence are

$$M_{iL} = M_i^* - K_C\{v[M_i^*]\}^{\frac{1}{2}} \quad \text{and} \quad M_{iU} = M_i^* + K_C\{v[M_i^*]\}^{\frac{1}{2}},$$

where K_C is the $\frac{1}{2}(100+C)$th standard normal percentile. As in Chapter 4, this naive interval is too short. Thus the naive interval makes the MCF estimate look more accurate than it actually is. Of course, a poor confidence interval is better than none, provided that it is viewed skeptically. In product applications, these naive limits may be close to the Nelson limits, provided that the units are homogeneous with respect to materials, manufacture, operation, etc. In contrast, these naive limits for the tumor data are significantly narrower than Nelson's limits, because the patients differ with respect to tumor recurrence rates. These naive limits do not extend to cost or value data.

Steps. Instead of using the formulas above, one can systematically calculate the naive confidence limits with the following steps. As follows, add columns to the above table containing the MCF calculation. For illustration, columns 2, 6, and 7 from Table 5.1 for childbirths to male statisticians are reproduced in Table 5.2. Steps (a)–(e) are the same as those above (p. 81):

(f) *Variance increments*: Proceeding down column 8 of Table 5.2, successively calculate the variance estimates of the increment m_i as $v(m_i) = m_i/(N_i - 0.5C_i)$ for each interval i with recurrences. For example, for the 30–32 interval, $v(m_{32}) = 0.214/(53 - 0.5 \times 3) = 0.00415$. $v(m_i) = 0$ for any interval with no recurrences; this is indicated with zero in column 8.

(g) *Variances*: Set $v[M_0^*] = 0$. Proceeding down column 9, successively calculate the variance estimates of the MCF M_i^* as $v[M_i^*] = v[M_{i-1}^*] + v(m_i)$ for each interval i with recurrences. For example, for the 30–32 interval, $v[M_{32}^*] = 0.00626 + 0.00415 = 0.01041$.

(h) *Half-widths*: Proceeding down column 10, calculate the two-sided confidence interval half-width $K_C\{v[M_i^*]\}^{1/2}$ for each interval i with recurrences. Here K_C is the

5.3. Confidence Limits

Table 5.2. *Calculation of naive confidence limits for men (** indicates omitted data).*

2. Endpoints $t_{i-1} - t_i$	6. $m_i = R_i/(N_i - 0.5 \times C_i)$	7. $M_i^* = M_{i-1}^* + m_i$	8. Increment $v(m_i) = m_i/(N_i - 0.5C_i)$
0	0.000	0.000	0.00000
0–20	0.000	0.000	0.00000
20–22	0.051	0.051	0.00088
22–24	0.052	0.103	0.00089
24–26	0.052	0.155	0.00091
26–28	0.088	0.244	0.00157
28–30	0.110	0.354	0.00202
30–32	0.214	0.567	0.00415
32–34	0.126	0.694	0.00266
34–36	0.205	0.898	0.00465
36–38	0.145	1.043	0.00348
38–40	0.052	1.095	0.00135
40–45	0.209	1.304	0.00624
45–50	0.087	1.391	0.00378
50–55	0.000	1.391	0.00000
55–60	0.000	1.391	0.00000
60–99	**	**	**

9. $v(M_i^*) = v(M_{i-1}^*) + v(m_i)$	10. $1.960 \times [v(M_i^*)]^{1/2}$	11. Lower limit	12. Upper limit
0.00000	0.0000	0.000	0.000
0.00000	0.0000	0.000	0.000
0.00088	0.0580	−0.007	0.109
0.00177	0.0824	0.021	0.185
0.00268	0.1014	0.054	0.257
0.00424	0.1277	0.116	0.371
0.00626	0.1551	0.199	0.509
0.01041	0.2000	0.367	0.767
0.01307	0.2241	0.470	0.918
0.01772	0.2609	0.637	1.159
0.02120	0.2854	0.757	1.328
0.02255	0.2943	0.800	1.389
0.02879	0.3326	0.971	1.636
0.03257	0.3537	1.037	1.744
0.03257	0.3537	1.037	1.744
0.03257	0.3537	1.037	1.744
**	**	**	**

$\frac{1}{2}(100 + C)$th standard normal percentile; for example, $K_{95} = 1.960$. Then for the 30–32 interval, the half-width of the two-sided approximate 95% confidence interval is $1.960\{0.01041\}^{1/2} = 0.2000$.

(i) *Limits*: Proceeding down column 11, calculate the lower limit for $M(t_i)$ as $M_i^* - K_C\{v[M_i^*]\}^{1/2}$ for each interval i with recurrences. For example, for the 30–32 interval, the lower limit is $0.567 - 0.2000 = 0.367$. Similarly, proceeding down column 12, calculate the upper limit for $M(t_i)$ as $M_i^* + K_C\{v[M_i^*]\}^{1/2}$ for each i. For example, for the 30–32 interval, the upper limit is $0.567 + 0.2000 = 0.767$.

(j) *Plot*: For each interval i, plot the lower and upper limits on the MCF plot at the upper endpoint t_i of the interval. Figure 5.2 is such an MCF plot with naive two-sided confidence limits for the births MCF for men. A pair of limits extends through intervals

86   Chapter 5. MCF Estimate and Limits for Interval Age Data

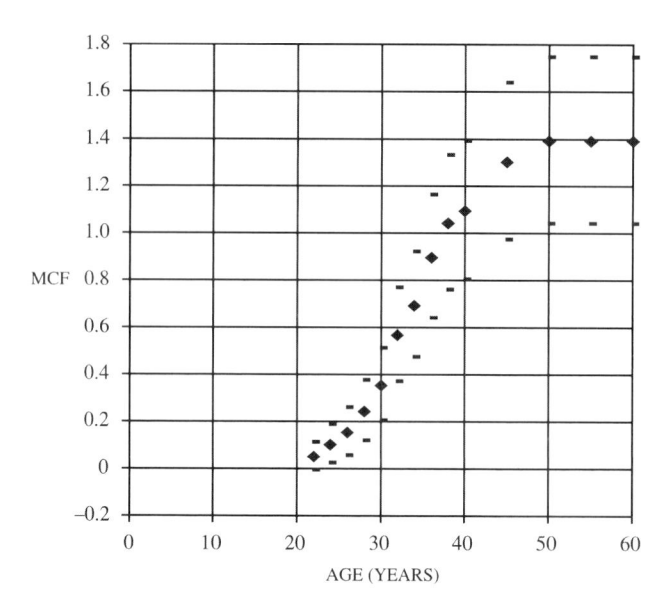

Figure 5.2. *Births MCF for men and naive 95% limits.*

to their right that have no recurrences and up to the next interval that has recurrences. The last pair of limits extends to the longest sample age, which is above 60 years.

As before, the calculations for the last interval are omitted since no units went completely through the interval.

Interpretation. An important feature of the confidence limits for births is their ultimate values. The estimate of the ultimate mean cumulative number of births per male statistician is 1.39, and the corresponding 95% limits are ±0.35. These limits are well below the 2.1 births needed to sustain a population. Thus there is convincing evidence that male statisticians average well below 2.1 births.

Software. The SAS Institute has developed software that calculates the preceding sample MCF and naive limits from interval age data. SAS will release this in 2002 as part of the Reliability Procedure of the SAS/QC software. Also, in 2002, ReliaSoft Corporation plans to develop such features in the recurrence data add-on to Weibull++6. In addition, the estimate and limits are easy to calculate and plot with a spreadsheet. Table 5.2 and Figure 5.2 were generated with Excel.

Assumptions. Assumptions $\langle 1 \rangle$–$\langle 8' \rangle$ above for the MCF estimate must hold. In addition, the naive limits require the following additional assumptions:

$\langle 9' \rangle$ The count data are from a nonhomogeneous Poisson process (Chapter 8) with MCF $M(t)$. This assumption is often dubious and means the following:

$\langle 9'a \rangle$ The process has independent increments. That is, any nonoverlapping age intervals have statistically independent numbers of recurrences.

$\langle 9'b \rangle$ The number of recurrences in any interval $(t, t']$ has a Poisson distribution (see Chapter 8) with mean $M(t') - M(t)$.

$\langle 10' \rangle$ The sampling distribution of $M_i^* = M^*(t_i)$ is close to normal. This is satisfied less well at the extremes of the age range. At low ages with few accumulated recurrences, the Poisson limits (Chapter 4) are often better.

⟨11'⟩ The adjustment $-0.5C_i$ in the variance increment properly adjusts for the unknown exact censoring ages in interval i.

In practice, the effect of departures from these assumptions must be evaluated to assess the adequacy of the naive limits. For example, when no units go entirely through the last interval I, the adjustment ⟨11'⟩ yields a denominator $(N_i - 0.5 \times C_I)$ that is biased high, and the variance increment is underestimated and is best omitted.

Problems

5.1. Births to women. The data on childbirths to female statisticians appear in Table 5.3:

(a) By hand or using a spreadsheet, calculate the sample MCF for women.
(b) Plot the sample MCF on Figure 5.1 to verify the plot.
(c) Calculate the naive two-sided 95% confidence limits for the MCF.
(d) Plot these limits on Figure 5.1.

5.2. Defrost control. The following application is a favorite of the author, as it dramatically illustrates some fundamental principles of statistics. Table 5.4 below shows field data (grouped by months in service) on replacements of defrost controls in refrigerators. A key engineer had made plots of the sample MCF on linear and log-log grids. Extrapolation of his plots yielded an estimate that 300% of the controls (three per refrigerator) would be replaced over a 15-year typical life of such refrigerators.

In view of this plot, the general managers and engineers of the refrigerator and appliance control departments were meeting weekly to make plans to redesign the control and to increase production to meet the projected demand for replacements. The author was asked to analyze the data and show that there was no problem.

Solve the problems below before reading the discussion that follows them. In the data table, treat the "Number at risk" as the number entering the month. The number censored in each month was not given. Do the problems by hand or with a spreadsheet or computer program.

(a) Determine appropriate numbers censored C_i in each month, especially months 12 and 24, and state your justification for your numbers.
(b) Calculate the MCF estimate for the cumulative percent replaced, using your C_i's. The engineer left out the $-0.5C_i$ adjustment in his calculation of the sample MCF. Does the engineer's MCF differ significantly from your MCF?
(c) Plot your MCF on linear and log-log grids. Extrapolate the sample MCF to estimate the cumulative percent replaced within a typical refrigerator life of 15 years (180 months).
(d) Calculate the naive approximate 95% confidence limits for the MCF. Put them on your plots.

Discussion. Initial discussions with the key engineer yielded the following information. (As is typical, much of it is irrelevant.) Historically, replacement rates on appliance controls varied considerably from market to market, and New York City's rate was highest—triple the

Table 5.3. *Data on childbirths to female statisticians.*

1. Interval i	2. Endpoints $t_{i-1} - t_i$	3. Number of recurrences R_i	4. Number censored C_i	5. Enter $N_i = N_{i-1} - C_{i-1}$	6. Increment $m_i = R_i/(N_i - 0.5 \times C_i)$	7. $M_i^* = M_{i-1}^* + m_i$
0	0					
1	0–20					
2	20–22					
3	22–24	1	1			
4	24–26	6	2			
5	26–28	8	7			
6	28–30	6	3			
7	30–32	9	3			
8	32–34	7	3			
9	34–36	3	2			
10	36–38	5	1			
11	38–40		1			
12	40–45		9			
13	45–50		3			
14	50–55		4			
15	55–60		4			
16	60–99		2			
	Total:	45	45			

Table 5.4. *Field data on replacements of defrost controls in refrigerators (grouped by months in service).*

Month i	Number replaced R_i	Number at risk N_i	Number censored C_i	Increment $m_i = R_i/(N_i - 0.5 \times C_i)$	MCF M_i^*
1	83	22914			
2	35	22914			
3	23	22914			
4	15	22914			
5	22	22914			
6	16	22914			
7	13	22911			
8	12	22875			
9	15	22851			
10	15	22822			
11	24	22785			
12	12	22745			
13	7	2704			
14	11	2690			
15	15	2673			
16	6	2660			
17	8	2632			
18	9	2610			
19	9	2583			
20	5	2519			
21	6	2425			
22	6	2306			
23	6	2188			
24	5	2050			
25	7	862			
26	5	845			
27	5	817			
28	6	718			
29	3	590			

national rate. Inspection of replaced controls returned to the factory showed that a significant percentage of them worked. Repair people possibly were "shotgunning" some refrigerators, replacing components, some of them working, until a problem was corrected. Therefore, we decided to call the repair a replacement rather than a failure. Also, we decided to count all replacements, rather than just true failures, as every replacement incurred a cost.

Whether or not the $-0.5C_i$ adjustment is used, the plots are virtually the same here. Examination of the MCF plots on square and log-log grids reveals a sharp increase in the MCF slope starting at 12 months and another at 24 months. When asked what is happening at 12 months, the engineer explained that the control fails when silicon lubricant seeps out of a bearing and penetrates the insulation, which then fails electrically, this process taking 12 months, which also happens to be the end of warranty. This physical explanation was hard to believe, as nature does not produce such sharp discontinuities in product behavior. Moreover, many refrigerators spent months in warehouses before installation, during which the lubricant seeped, and storage would smear any sharp increase of the MCF out over several months.

When asked why the numbers of refrigerators observed in months 13 through 24 were so much smaller than the numbers observed in months 1 through 12 (and yet smaller in months

25 through 29), the engineer said the first twelve months of data came from refrigerators whose owners had sent in a dated purchase-record card (about 65% of the population) and that thus had a known installation date. This date and the date of a control replacement were used to calculate the refrigerator's age at each replacement. He said the data on months 13 through 24 came from refrigerators (about 7% of the population) whose owners extended the warranty for a year by purchasing a one-year service contract for all parts and labor. The linear plot shows that this group had a replacement rate about three times that of the 65% of the population. Similarly, he said that data on months 25 through 29 came from refrigerators (about 2% of the population) whose owners bought a second year of service contract. The data show that these refrigerators had a replacement rate about three times that of refrigerators on service contract the previous year. Within each of the three years, the replacement rate was essentially constant. Who buys a service contract and who renews it? It is likely to be those who had repair problems while under warranty. Thus the second and third years of data are representative of subpopulations with high replacement rates, not the population as a whole. Consequently, the MCF plot consists of pieces of the MCFs of three subpopulations (65%, 7%, and 2%), giving the appearance of an increasing replacement rate (derivative) where, in reality, the population rate was essentially constant.

As the first year of data has 65% of the refrigerators, it is most representative of the entire population. Linear extrapolation of the MCF of just the first year of data yields an estimate of about 6% replacement over typical refrigerator life (180 months), not 300%. This explanation convinced management that there was no problem. Later experience bore this out.

This application illustrates two important statistical principles:

1. Data are representative of only the part of the population they come from. Here the data were *not* a random sample from the target population (the entire population). Instead the second and third years of data were self-selected samples with high replacement rates. This underscores the importance of having a truly random sample.

2. Data plots are essential for understanding data. No fitting of mathematical models to the data for this application could have led to correct understanding and conclusions.

5.3. Tumor data. Analyze the placebo data of Table 1.2, treating the data as interval data grouped by month.

(a) Calculate the MCF estimate.

(b) Plot the estimate on Figure 4.3(a). Compare this estimate with the previous one.

(c) Calculate the naive 95% confidence limits for the MCF.

(d) Plot the naive limits on Figure 4.3(a). Compare these limits with the Nelson limits in the plot.

(e) Repeat (a)–(e) for the Thiotepa data in Problem 3.2 and Figure 4.3(b).

5.4. Traction motors. The traction motor data (Table 1.6) are interval data, which are analyzed in Chapter 6. See Problem 6.1, which requires interval data analyses.

5.5. "Exact" birth data. Reanalyze the births data for male statisticians. For each age interval, treat the births as occurring at distinct ages spread evenly over the interval.

Problems   91

(a) Estimate the MCF. Plot it on Figure 5.2. Describe how the two estimates compare. Which do you prefer and why?

(b) Calculate the corresponding 95% confidence limits for exact data. Plot the limits on Figure 5.2. Describe how the two sets of limits compare. Which do you prefer and why?

5.6. Gearboxes. Table 5.5 shows dealer claims on replacements of power steering gearboxes in a fleet of 55,290 cars of a recent model. The data are collected to monitor and forecast claims. The table heading shows the sale month and the number sold that month; for example, 3834 cars were sold in month 1, and they are treated as a group (or subpopulation). The first column shows the age of cars (months in use since sale). The body of the table shows the *number* of claims for each combination of sale month and age; for example, among the 4486 cars sold in month 2, there were two claims in month 1 in use and two claims in month 2 in use. The ends of history for the cars in a sale month are spread over the final month in that column; for example, for sales month 1, the censoring ages of the 3834 cars are spread over month 38. For each month of sale, it is common practice to ignore the last month of use, since claims that month are often underreported due to delays in dealers submitting claims. Lawless (1995b) and Kalbfleisch, Lawless, and Robinson (1991) correct for such reporting delays.

(a) Month of sale for cars corresponds roughly to month of production. Most production processes continue to be improved after startup, and thus early months of production have more failures than later months. For selected early and late months of sale, calculate and plot each month's sample MCF for the number of claims. Decide whether to plot a point for each claim or for each month, and explain your choice. From your plots, judge whether early sales months have higher claim rates. Explain your reasoning.

(b) Use the totals in the right-hand column to get a pooled estimate of the fleet MCF for the number of claims. Plot this estimate on a linear grid and on a log-log grid. Interpret the plots. In particular, does the claim rate increase or decrease as the population ages?

(c) Using a plot, estimate the claim rate at 48 months, the end of the warranty.

(d) Criticize this pooled estimate (see Figure 3.6). Explain what is needed to improve the population MCF estimate.

(e) To predict future claims, would you prefer to use the pooled MCF or a separate MCF for each sales month? Why?

Note that confidence limits for the MCF are not appropriate here, as the sample *is* the population. Prediction limits like those of Robinson (1995) are more appropriate. His limits for exact age data need to be extended to interval age data.

For the following analyses, use the cost data in Table 5.6. The entry in the table is the *total dollar cost* of all claims for that sale month and month in use. For example, for sale month 2, the total cost of the two claims in month 2 in use is $1896. Costs are rounded to the nearest dollar; thus some row sums appear off by a dollar.

(f) For selected early and late months of sale, calculate and plot each month's sample MCF for the cost of claims. From your plots, judge whether early sales months have higher claim costs. Explain your reasoning.

Table 5.5. *Gearbox replacements versus sale month and month in use.*

Month in use	\multicolumn{16}{c	}{Sale month and number sold}	Row sum
	1	2	3	4	5	6	7	8	9	10	11	12	13	14	15	
	3834	4486	4085	3599	3289	4157	3879	4673	4981	4564	3631	6136	3404	491	81	55290
1	0	2	2	0	2	0	0	0	0	1	1	0	0	0	0	8
2	0	2	0	0	0	0	0	2	1	0	0	0	0	0	0	5
3	1	0	0	0	1	0	0	1	1	0	0	0	0	0	0	4
4	0	0	1	1	0	0	0	0	0	0	0	1	0	0	0	3
5	2	2	1	0	0	0	0	0	1	0	1	0	0	0	0	7
6	1	2	3	0	0	0	0	0	0	0	0	1	2	0	0	9
7	1	0	1	0	0	0	0	1	0	1	0	0	1	0	1	7
8	1	1	0	1	0	1	0	0	2	1	0	1	0	0	0	7
9	1	2	0	0	0	0	0	0	0	0	0	1	1	0	0	5
10	1	0	1	0	1	0	1	1	0	0	0	0	0	1	0	6
11	0	1	1	1	0	1	0	0	0	2	0	0	0	0	0	5
12	2	2	0	1	0	0	0	0	0	0	0	0	2	0	0	6
13	1	1	2	1	0	0	0	0	0	1	2	1	1	0	1	10
14	0	2	0	0	0	0	0	2	0	1	0	0	0	0	0	5
15	1	0	0	0	0	0	0	1	0	1	1	2	1	0	0	7
16	2	1	0	0	0	0	0	0	0	1	2	3	2	1	0	13
17	0	3	1	0	0	0	0	0	2	1	0	0	0	0	0	7
18	1	1	1	0	0	0	1	0	0	0	0	0	1	0	0	6
19	0	2	1	0	0	0	0	0	0	2	2	0	0	0	0	7
20	0	0	0	1	1	2	0	0	1	1	0	1	1	0	0	9
21	1	3	1	2	0	0	0	0	0	1	0	1	1	0	0	10
22	0	0	0	0	0	1	0	0	1	0	1	0	0	0	0	5
23	1	1	0	0	0	0	0	1	1	0	0	1	0	0	0	5
24	0	0	0	1	0	2	0	0	2	0	0	1	0	0	0	6
25	2	3	0	0	0	0	0	0	0	0	1	1	0	0		7
26	2	1	0	1	0	0	0	0	3	1	0	0	0			8
27	2	1	1	0	1	0	0	1	0	0	0	0				5
28	0	0	0	0	0	0	0	0	2	0	0					3
29	0	1	0	0	1	0	3	1	1	0						7
30	1	1	0	1	0	0	0	0	0							2
31	0	0	0	0	0	0	0	0								2
32	0	0	0	0	0	3	0									3
33	0	2	1	0	0	0										3
34	1	4	0	0	0											5
35	0	2	1	0												3
36	1	1	1													3
37	1	0														1
38	0															0
Sum:	28	45	21	10	8	10	5	11	17	16	12	14	13	2	2	214

Table 5.6. *Gearbox replacement costs ($) versus sale month and month in use.*

Month in use	\multicolumn{15}{c}{Sale month and number of cars sold}	Row sum
	1	2	3	4	5	6	7	8	9	10	11	12	13	14	15	
	3834	4486	4085	3599	3289	4157	3879	4673	4981	4564	3631	6136	3404	491	81	54718
1		1643	1911		1669					1000	772					6994
2		1896						1726			771					4394
3	864				765			768	822							3219
4	767		725	730					856							3078
5	720	1730	1096									1020	878			5445
6	1294	1485	2444										790	1735		7749
7	784		1287					856		810	868		760		819	6183
8	738	761		828		786			1786	887						5787
9	998	1548										815	763			4124
10	853		804		949		857	829						915		5207
11		738	625	779						1568						3711
12	1773	1484		800		746										4804
13	761	768	1854		1420							1299		1175	556	7835
14		1632						778		729			602	582		4323
15	1041							689		662			1517	829		4738
16	2824	880						620		749	1197	2282	1163	582		10296
17		2452	1177						1272	591						5492
18	818	1169	840				916			559			705			5007
19		1581	624							1164	1117	541				5026
20				656	639	2628			634	550	559		554			6221
21	1206	2390	565	1373						571			571	630		7307
22	603		624			578			582		640					3027
23	769	595					691	560				573				3189
24				563		1393		1187				591				3735
25	1255	1609									578	573				4015
26	1548	729		610					1666	559						5113
27	1240	581	563					571								2955
28									1499							1499
29		661			618		2106	624	794							4803
30	568	613														1181
31		548		558	467											1573
32						1930										1930
33		1173	596													1769
34	572	3218														3790
35			597	646												1243
36	611	569	624													1804
37	577															577
38																

(g) Use the row totals in the right-hand column to get a pooled estimate of the fleet MCF for the cost of claims. Plot this estimate on a linear grid and on a log-log grid. Interpret the plots. In particular, does the cost per month per car increase or decrease as the population ages?

(h) Using a plot, estimate the mean cumulative claim cost at 48 months, the end of the warranty.

(i) Criticize this pooled estimate (see Figure 3.6). Explain what is needed to improve the population MCF estimate for cost.

(j) To predict future costs of claims, would you prefer to use the pooled MCF or a separate MCF for each sales month? Why?

For some work, people prefer to use mileage instead of months in use. Then the current mileages of all cars in the fleet are needed to calculate a sample MCF for the fleet. Usually these mileages are not known and must be estimated. For each of the 209 replacements of original gearboxes, Table 5.7 shows the car's number of days in service and corresponding mileage. The five repeat gearbox replacements are not included in this table. The total number of days for this table is 103,085 days, and the total mileage is 4,256,609. The number of miles per day was calculated for each car except the first one, and their average miles per day is 42.33. All sales months in previous tables contain exactly 30 days.

Table 5.7. *Days in service and mileage when each gearbox was replaced.*

Days	Miles	Days	Miles	Days	Miles	Days	Miles	Days	Miles
0	26	197	3885	409	16508	583	16495	761	34586
8	25	214	10082	409	41218	587	15193	764	35917
8	293	217	11440	419	16469	594	13230	765	67040
10	1561	220	8256	419	34357	594	15000	776	6306
11	737	224	13600	421	9825	595	27054	781	38828
14	45	231	9924	423	13671	596	37833	784	50009
16	484	236	6102	434	12381	597	47942	796	34737
26	240	237	12521	441	8829	605	37108	797	30362
34	942	242	3617	444	23435	609	27169	805	20050
42	1658	243	10086	448	12321	611	45928	831	32293
44	1507	251	10132	455	18764	612	24618	839	24183
48	1215	257	23919	457	3402	617	10500	847	24356
55	2376	259	9736	458	32940	619	31100	850	24913
61	3068	275	11547	459	11142	620	17730	852	35398
69	2639	279	21629	460	10022	620	41201	853	28637
70	2680	284	13487	462	7380	623	45407	859	33545
78	3112	292	13085	464	11425	625	37698	859	32303
101	1845	293	22875	466	9962	639	31122	862	14972
103	3669	295	12120	467	11093	648	23791	877	48188
116	6900	303	4726	477	34418	657	17501	883	27507
120	11133	308	12039	477	14845	658	21843	907	26703
127	3047	314	14100	478	13396	665	29056	916	48364
130	7815	320	11290	479	24762	667	25446	924	59751
141	3696	322	16455	485	27884	672	33408	949	24864
147	2712	338	8389	485	33866	685	26193	950	45864
149	8133	338	7400	489	24314	685	18546	952	30074
149	6546	347	13726	500	31210	693	27620	970	58010
153	2603	350	16577	504	23485	694	20581	971	47865
153	5856	351	9993	518	12568	697	22763	985	41882
156	5696	353	14510	520	23268	697	28355	994	30659
158	43202	362	13550	522	10689	707	23301	996	54087
158	3901	367	12645	529	17804	710	24951	999	29548
163	10000	374	15279	529	12044	721	18520	1008	33856
163	5151	374	16419	539	42667	723	24583	1012	44521
166	7586	378	15500	545	27055	725	43998	1020	24663
178	2743	383	8117	547	30210	728	32161	1026	21464
181	9012	385	20266	552	27644	739	22960	1042	45208
185	6982	386	8930	560	18300	742	14979	1061	49072
186	6564	387	12224	562	24917	745	13501	1062	20968
188	5590	388	18298	565	31692	754	30845	1078	42235
195	7163	398	27249	570	40507	756	48456	1092	30000
195	6474	406	26257	580	23650	756	52187		

(k) Use the table to estimate the distribution of current (censoring) mileages for the fleet of 55,290 cars.

(l) Criticize your distribution estimate and state its biases.

(m) For selected early and late months of sale, calculate and plot each month's sample MCF for the *number* of claims as a function of mileage. Decide whether to plot a point for each claim or for each month, and explain your choice. From your plots, judge whether early sales months have higher claim rates as a function of mileage. Explain your reasoning.

Problems

(n) Use the totals in the right-hand column of Table 5.5 to get a pooled estimate of the fleet MCF for the number of claims as a function of mileage. Plot this estimate on a linear grid and on a log-log grid. Interpret the plots. In particular, does the population claim rate increase or decrease with mileage?

(o) Using a plot, estimate the claim rate at 50,000 miles, the end of the warranty.

(p) Criticize this pooled estimate (see Figure 3.6). Explain what is needed to improve the population MCF estimate.

(q) For selected early and late months of sale, calculate and plot each month's sample MCF for the cost of claims as a function of mileage. From your plots, judge whether early sales months have higher claim costs. Explain your reasoning.

(r) Use the totals in the right-hand column of Table 5.6 to get a pooled estimate of the fleet MCF for the cost of claims as a function of mileage. Plot this estimate on a linear grid and on a log-log grid. Interpret the plots. In particular, does the cost per mile per car increase or decrease with mileage?

(s) Using a plot, estimate the mean cumulative claim cost per car at 50,000 miles, the end of the warranty.

(t) Criticize this pooled estimate (see Figure 3.6). Explain what is needed to improve the population MCF estimate for cost as a function of mileage.

Chapter 6
Analysis of a Mix of Events

6.1 Introduction

Purpose. This chapter presents methods for analyzing recurrence data with a mix of events/recurrences. For example, for the childbirth data (Table 1.5), two distinct events are boy and girl. Many data sets contain more than one type of recurrence. For example, data on products may contain any number of failure modes. Indeed, the naval turbine data of Chapters 1 and 2 contain a large number of failure modes. Few applications have just one type of recurrence.

Data. The data on traction motors for subway cars (Table 1.6) are used here to illustrate such analyses. This data set on $N = 372$ cars, each with four motors, has $K = 21$ failure modes (A, B, C, etc.) and 164 failures. Information was sought for two main purposes: (1) to evaluate the probability that a new fleet for a customer would pass a reliability demonstration test in service and (2) to identify how best to improve the current design. In particular, the manufacturer sought the MCF estimate for the following:

- All failure modes combined (section 6.2). This yields information on the probability of passing the demonstration test.

- Each failure mode separately (section 6.3). This shows how much each mode contributes to the total MCF and identifies which failure modes warrant improvement.

- Groups of failure modes (section 6.4). The failure modes were divided into three groups: modes in (due to) the design, nondesign modes, and an unassigned mode K. The MCF estimates indicate where effort is needed.

- Modes eliminated (section 6.5). It was thought that modes A, B, and C could be eliminated through redesign. Would the reduction in the MCF be worthwhile?

Analyses that yield this information follow. The application required many analyses beyond those shown here.

Overview. This chapter contains the following topics:

6.2 *Model for a mix of events*: This section describes the population model in which each population unit has a cumulative history function for each type of event. Then each type of event has an MCF.

6.3 *MCF with all types of events combined*: This section shows how to estimate the population MCF for all types of events combined, always of interest.

6.4 *MCF for a single type of event*: This section shows how to estimate the population MCF for a single type of event, for example, birth of a boy. For the traction motors, the cause of failure mode K was not understood, and its MCF was sought to gain insight into its cause.

6.5 *MCF for a group of events*: This shows how to estimate the population MCF for a group of events, for example, all failure modes of a particular component or subsystem. This MCF shows how much the component contributes to the system MCF.

6.6 *MCF with events eliminated*: Products can be redesigned to eliminate failure modes. This section shows how to estimate from existing data the MCF that would result if those modes were eliminated. For the traction motors, engineers proposed to eliminate modes A, B, and C and wanted to know how that would affect the resulting population MCF.

6.7 *Practical and theoretical issues*: This section discusses the issues that must be properly dealt with to get reliable results from the methods of this chapter.

Problems.

Software. The MCF calculations and plots below can be generated for exact age data with right censoring with the software described in Chapters 3 and 4. Also, they can be made with a spreadsheet. The following calculations and plots for interval age data were made with Excel. The only commercial software that calculates the MCF estimate and naive limits for interval age data will be released by the SAS Institute in 2002 as part of the Reliability Procedure of SAS/QC software.

6.2 Model for a Mix of Events

Purpose. This section describes a nonparametric model for a population with a mix of types of events/recurrences.

Types of events. All of a population's possible events or recurrences must be partitioned into distinct types. That is, each recurrence must be assigned to one type of event. This requires a clear operational definition of each type. The choice of partition depends on the application, and different partitions can be used for difference purposes with the same data set. For example, childbirths could be partitioned into (a) boy and girl; (b) live and stillborn; (c) caesarian and natural; (d) black, white, Asian, Indian, etc.; (e) single, twins, triplets, etc.; (f) and so on. Similarly, product failures could be partitioned by cause, failure mechanism, subsystem, component, symptom, consequences, etc. Hereafter, we assume that there are K distinct types of events and number them $k = 1, 2, \ldots, K$.

History functions. In previous chapters, each population unit n has a single cumulative history function $Y_n(t)$. Here population unit n has a cumulative history function for each type k of event, namely, $Y_{n1}(t), Y_{n2}(t), \ldots, Y_{nK}(t)$, a vector of functions, variables, or curves. For example, each statistician has a cumulative history function for the number of boy births and another for the number of girl births. The "data value" for a unit is K-variate. But instead of K numbers, each unit has K functions. The K functions can be staircase functions or continuous, staircases can have steps of size 1 (a count of recurrences) or steps of any size (cost or value), and functions can increase or decrease.

6.2. Model for a Mix of Events

Model. The nonparametric model for a population of N units is the population of the N vectors, each with K cumulative history functions, one for each type of event. That is, the model is the population of N uncensored vectors $[Y_{n1}(t), Y_{n2}(t), \ldots, Y_{nK}(t)]$, each with K curves, $n = 1, 2, \ldots, N$. Alternatively, one can view each type k of event as a separate population of N curves $Y_{nK}(t), n = 1, 2, \ldots, N$. The correlations and joint structure among the K types of events are not discussed here. This model is a simple representation of a multivariate stochastic process.

Distribution. At a particular age t, the corresponding values of the N vectors $[Y_{n1}(t), Y_{n2}(t'), \ldots, Y_{nK}(t)]$ have a K-variate joint distribution. This joint distribution is a function of t. When the $Y_{nk}(t)$ are counts of recurrences, this joint distribution is discrete. For some cost or value variables $Y_{nk}(t)$, the joint distribution may be regarded as continuous.

Censoring. Age data on sample history functions may have right, left, or gap censoring. Censoring is a property of the data and its collection, not the model. As before, the population history functions in the model have no censoring.

MCFs. Each type k of event has a population MCF function $M_k(t), k = 1, 2, \ldots, K$. At age t', by definition, $M_k(t') \equiv E_n\{Y_{nk}(t')\}$ is the expectation (average) of all N population values $Y_{nk}(t'), n = 1, 2, \ldots, N$. Thus the population has a K-variate mean curve $[M_1(t), M_2(t), \ldots, M_K(t)]$.

MCFs add. Often one is interested in the sum of the costs or number of recurrences of a group G of types of events. For example, a group of failure modes of the traction motors are attributed to design. The MCF for the group is the sum of the MCFs of those events; namely,

$$M_G(t) = \sum_{k \text{ in } G} M_k(t). \tag{6.1}$$

That is, MCFs add—an obvious property. In particular, this relationship holds when G is the group of all types of events. This relationship is used repeatedly for analysis of data with a mix of types of events. This relationship is a consequence of the fact that the MCF is an expectation (mean) over all N population units. Namely,

$$M_G(t) = E_n Y_{nG}(t) = E_n \left\{ \sum_{k \text{ in } G} Y_{nk}(t) \right\} = \sum_{k \text{ in } G} E_n\{Y_{nk}(t)\} = \sum_{k \text{ in } G} M_k(t).$$

Rate. As before, in some applications, $M_k(t)$ is assumed to have a derivative

$$m_k(t) = \frac{dM_k(t)}{dt}. \tag{6.2}$$

This mean recurrence rate or cost rate for event type k applies to the number or cost (value) of recurrences, as before. Similarly, the rate $m_G(t)$ for a group G of types of events is

$$m_G(t) = \sum_{k \text{ in } G} m_k(t). \tag{6.3}$$

That is, rates add. In particular, this relationship holds when G is the group of all types of events.

Assumptions. The model assumptions of Chapter 2 apply to the model here for a mix of types of events. In practice, it is necessary to assess assumptions to ensure that the results of analyses are reliable. Such issues are discussed in section 6.7. As in previous chapters, parametric stochastic process models and their assumptions are not used here.

Life data. The model for the *life* of a series system with a mix of independent competing failure modes is widely used in reliability work. Nelson (1982, Chapter 5), Meeker and Escobar (1998), Crowder (2001), and David and Moeschberger (1978) describe it in detail. It has little relation to the model here for a mix of types of repeated events.

6.3 MCF with All Types of Events Combined

Purpose. This section shows how to estimate the population MCF with all K types of events combined, namely, $M_{\text{all}}(t) = M_1(t) + M_2(t) + \cdots + M_K(t)$. This MCF serves as a base for seeing how much the different types of events contribute to it.

MCF. The sample MCF with all types of events combined is calculated as follows. Combine all data, ignoring types of events. This is done in Table 6.1 for the traction motor data; the second column contains the total number of failures from all modes observed in each month. (Each monthly total is obtained by summing across all modes in that month's row of Table 1.6.) The MCF estimate (expressed as a percentage) in Table 6.1 is calculated as described in Chapter 5 on interval data. Note that calculations are accurate to seven

Table 6.1. *MCF calculation for all modes combined.*

Month i	Number of all modes	Number enter month	Number censored	Increment % m_i	MCF % M_i^*
1	3	372		0.81	0.8
2	2	372		0.54	1.3
3	2	372		0.54	1.9
4	5	372		1.34	3.2
5	2	372		0.54	3.8
6	3	372		0.81	4.6
7	6	372		1.61	6.2
8	6	372	13	1.64	7.8
9	4	359	28	1.16	9.0
10	7	331	28	2.21	11.2
11	6	303	29	2.08	13.3
12	5	274	27	1.92	15.2
13	12	247	25	5.12	20.3
14	4	222	24	1.90	22.2
15	5	198	1	2.53	24.7
16	13	197	11	6.79	31.5
17	9	186	7	4.93	36.5
18	8	179	12	4.62	41.1
19	11	167	14	6.88	48.0
20	6	153	12	4.08	52.0
21	3	141	14	2.24	54.3
22	6	127	11	4.94	59.2
23	3	116	13	2.74	62.0
24	13	103	9	13.20	75.2
25	9	94	8	10.00	85.2
26	3	86	12	3.75	88.9
27	5	74	13	7.41	96.3
28	2	61	11	3.60	99.9
29		50	9		
30	1	41	13	2.90	102.8
31		28	13		
32		15	12		
33		3	3		

6.3. MCF with All Types of Events Combined

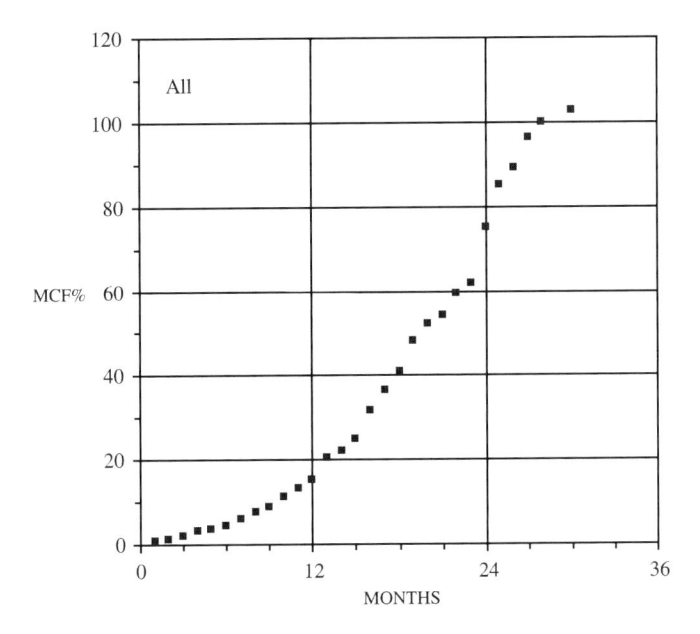

Figure 6.1. *MCF for all failure modes combined.*

figures and results are rounded to three; this accounts for small discrepancies in the table. Figure 6.1 shows the MCF plot. Here the MCF is the mean number of failures per car. As there are four motors per car, this MCF divided by 4 yields the MCF for motors. That is, $M_{car}(t) = 4 \times M_{motor}(t)$.

Interpretation. Figure 6.1 shows that the recurrence rate (derivative) for all modes initially increases and then becomes essentially constant after one year. The constant rate is about 66% (0.66 failures) per car per year. This MCF was used to estimate the chance of a new fleet passing a demonstration test in service. The data extend to 30 months, where the MCF estimate is 103% (1.03 failures) per car. Below, the various MCF estimates at 30 months are compared to assess how much each mode contributes to the total. In many applications, engineers use the end of warranty or design life for such comparisons.

Plots. Such a plot can be enhanced. As before, one can add confidence limits and display the censoring times. Also, the plotted points can be symbols that identify the types of events, for example, modes A, B, C, etc., for the traction motors. Such symbols reveal which types of events dominate as the population ages. Also, displays with such symbols (such as Figure 1.9) of individual unit histories on time lines may reveal useful information. This application has interval age data. Thus, as in Chapter 5, one could connect the plotted points with straight line segments, or one could draw or imagine a smooth curve through the points. Additionally, some analysts may prefer to plot a point for each recurrence and to spread the points evenly over their interval.

Extensions. The estimate extends to cost (or value) data, exact age data, age data with left censoring and gaps, and continuous history functions. Confidence limits can also be calculated and plotted. Since the data here are interval data, only the naive limits can be calculated. For each type of data, use the corresponding estimate from Chapter 3 and confidence limits from Chapter 4.

Number. Simply combining the *numbers* of all types of events may not be suitable in some applications. Types of events may differ significantly with respect to cost, value, importance, etc. Then it is better to use some suitable variable to put various events on a common scale.

6.4 MCF for a Single Type of Event

Purpose. This section shows how to estimate the MCF $M_k(t)$ for just a single type of event. Often one wishes to understand the behavior of that type of event or to compare the MCFs for different types of events.

Data. The data on failure mode K of the traction motors illustrate the calculation of the sample MCF for a single type of event. The cause of mode K ("unexplained flashover") was unknown. The engineers hoped that the MCF plot would provide some insight. In the actual application, the sample MCF was obtained for each individual mode with more than a few failures.

MCF. Calculate the sample MCF for a single type of event from the data in Table 6.2 as follows. Use all sample censoring ages, but use only the recurrence data on the selected type of event. The second column of Table 6.2 shows the number of mode K failures in each month. This is the column for mode K in Table 1.6; all other recurrences are ignored. The MCF estimate (expressed as a percentage) in Table 6.2 is calculated as described in Chapter 5 on interval data. Figure 6.2 shows its plot. Here and elsewhere, the MCF is the mean number of failures per car.

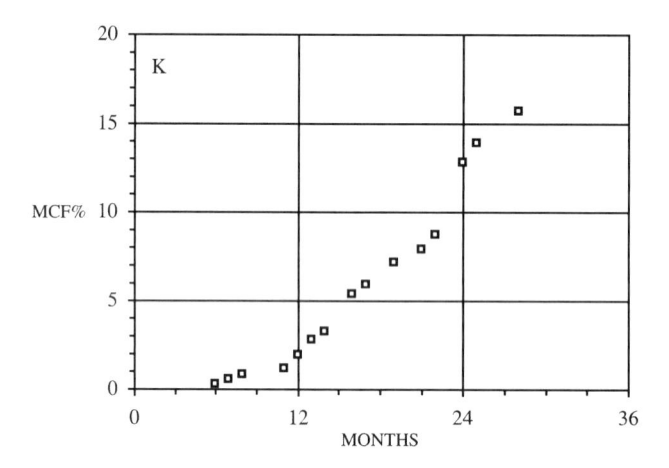

Figure 6.2. *MCF for failure mode K.*

Interpretation. Figure 6.2 shows that the recurrence rate for mode K initially increases up to the one year point and afterwards is relatively constant. The constant rate is about 11% (0.11 failures) per car per year. This is about 17% of the constant rate for all modes combined. Thus mode K is a major contributor to the total rate and interested the engineers. At 30 months, the MCF estimate is 16% (0.16 failures) per car, compared with 1.03 failures for all modes.

Extensions. As above, this MCF estimate extends to cost (or value) data, exact age data, age data with left censoring and gaps, and continuous history functions. Confidence limits can also be calculated and plotted. Since the data here are interval data, only the naive limits can be calculated. For each type of data, use the corresponding estimate from Chapter 3 and confidence limits from Chapter 4.

6.5. MCF for a Group of Events

Table 6.2. *MCF calculation for mode K.*

Month i	Number of mode K	Number enter month	Number censored	Increment % m_i	MCF % M_i^*
1		372			
2		372			
3		372			
4		372			
5		372			
6	1	372		0.27	0.27
7	1	372		0.27	0.54
8	1	372	13	0.27	0.81
9		359	28		
10		331	28		
11	1	303	29	0.35	1.16
12	2	274	27	0.77	1.93
13	2	247	25	0.85	2.78
14	1	222	24	0.48	3.25
15		198	1		
16	4	197	11	2.09	5.34
17	1	186	7	0.55	5.89
18		179	12		
19	2	167	14	1.25	7.14
20		153	12		
21	1	141	14	0.75	7.89
22	1	127	11	0.82	8.71
23		116	13		
24	4	103	9	4.06	12.77
25	1	94	8	1.11	13.88
26		86	12		
27		74	13		
28	1	61	11	1.80	15.68
29		50	9		
30		41	13		
31		28	13		
32		15	12		
33		3	3		

6.5 MCF for a Group of Events

Purpose. This section shows how to estimate the population MCF for a group G of types of events, namely, $M_G(t) = \sum_{k \text{ in } G} M_k(t)$. Such a group MCF is of interest in many applications. For example, in engineering, typical groups are all failure modes of a subsystem or of a component.

Data. The estimate of a group MCF is illustrated with two groups of traction motor failure modes. First is the design group (modes $A, B, C, D, E, J, M, N, P, Q, R, S$, and T). The design modes are due to the motor design, and their contribution to the total MCF is of interest. Second is the ABC group (design modes A, B, and C), which could be eliminated with a design improvement. Their contribution to the total must be known in order to judge whether eliminating them is worthwhile.

Design MCF. The sample MCF for a group G of types of events is calculated as follows. Use all sample censoring ages, but use only the recurrence data on the selected types of events in the group. Combine all data on those recurrences, ignoring all other recurrences. This is shown in Table 6.3 for the design modes. The second column contains the total number of failures from all design modes observed in each month. Each monthly total is obtained by summing across all design modes in that month's row of Table 1.6. The calculation of the MCF estimate (expressed as a percentage) in Table 6.3 is described in Chapter 5 on interval data. Figure 6.3 shows its plot.

Table 6.3. *MCF calculation for design modes.*

Month i	Number in design	Number enter month	Number censored	Increment % m_i	MCF % M_i^*
1	1	372		0.27	0.3
2	1	372		0.27	0.5
3	2	372		0.54	1.1
4	5	372		1.34	2.4
5	1	372		0.27	2.7
6	2	372		0.54	3.2
7	3	372		0.81	4.0
8	5	372	13	1.37	5.4
9	3	359	28	0.87	6.3
10	7	331	28	2.21	8.5
11	3	303	29	1.04	9.5
12	2	274	27	0.77	10.3
13	3	247	25	1.28	11.6
14	3	222	24	1.43	13.0
15	3	198	1	1.52	14.5
16	4	197	11	2.09	16.6
17	3	186	7	1.64	18.2
18	3	179	12	1.73	20.0
19	4	167	14	2.50	22.5
20	3	153	12	2.04	24.5
21	1	141	14	0.75	25.3
22	5	127	11	4.12	29.4
23	2	116	13	1.83	31.2
24	5	103	9	5.08	36.3
25	2	94	8	2.22	38.5
26	2	86	12	2.50	41.0
27	2	74	13	2.96	44.0
28	1	61	11	1.80	45.8
29		50	9		
30	1	41	13	2.90	48.7
31		28	13		
32		15	12		
33		3	3		

Design plot. Figure 6.3 shows that the recurrence rate (derivative) for design modes initially increases and then becomes essentially constant after one year, much like the rate for all modes. The constant rate is about 30% (0.3 failures) per car per year. The data extend to 30 months, where the MCF estimate is 49% (0.49 failures) per car. The corresponding MCF value for all modes is 1.03 failures per car. Thus design modes contribute about half of the motor failures. In many applications, engineers use the end of warranty or design life for such comparisons.

Enhancements. Such a plot can be enhanced. As before, one can add confidence limits and display the censoring times. Also, the plotted points can be symbols that identify the types of events, for example, design modes A, B, C, etc. This reveals which types of events dominate as the motors age.

6.6. MCF with Events Eliminated

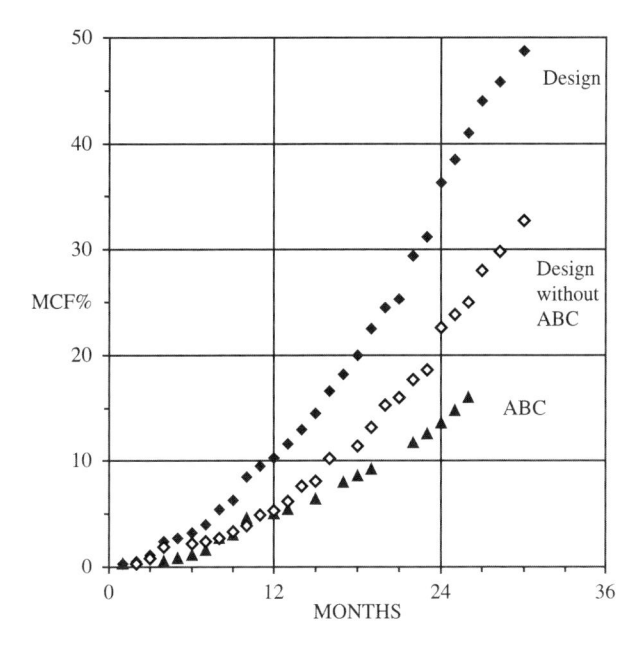

Figure 6.3. *MCF for ABC and design groups.*

***ABC* MCF.** The sample MCF for a group *ABC* is calculated in Table 6.4. There the second column contains the total number of failures from modes *A*, *B*, and *C* observed in each month. Each monthly total is obtained by summing the data across those three modes in that month's row of Table 1.6. The MCF calculation (as a percentage) in Table 6.4 is described in Chapter 5 on interval data. The MCF plot is in Figure 6.3.

***ABC* plot.** Figure 6.3 shows that the recurrence rate (derivative) for modes *ABC* gradually increases over the observed age range. At 30 months, the MCF estimate is 16% (0.16 failures) per car. The corresponding MCF value for design modes is 0.49 failures per car. Thus modes *ABC* contribute about one-third of the design failures. In many applications, engineers use the end of warranty or design life for such comparisons.

Combine plots. Figure 6.3 combines a number of MCF plots. This makes comparisons easier. Figure 6.3 compactly shows the design and *ABC* MCFs and the result of removing modes *A*, *B*, and *C*.

Extensions. As above, this MCF estimate extends to cost (or value) data, exact age data, age data with left censoring and gaps, and continuous history functions. Confidence limits can also be calculated and plotted. Since the data here are interval data, only the naive limits can be calculated. For each type of data, use the corresponding estimate from Chapter 3 and confidence limits from Chapter 4.

6.6 MCF with Events Eliminated

Purpose. This section shows how to estimate the population MCF that would result if certain types of events were eliminated. In engineering work, it is often possible to redesign the product to eliminate certain failure modes. Then one wants to estimate in advance the resulting MCF. This is a special case handled in the previous section, that of estimating the MCF for a group G of types of events, namely, $M_G(t) = \sum_{k \text{ in } G} M_k(t)$. Here the group G is all types minus the types to be eliminated.

Table 6.4. *MCF calculation for modes A, B, and C combined.*

Month i	Number of A, B, and C	Number enter month	Number censored	Increment % m_i	MCF % M_i^*
1	1	372		0.27	0.3
2		372			
3		372			
4	1	372		0.27	0.5
5	1	372		0.27	0.8
6	1	372		0.27	1.1
7	2	372		0.54	1.6
8	4	372	13	1.09	2.7
9	1	359	28	0.29	3.0
10	5	331	28	1.58	4.6
11		303	29		
12	1	274	27	0.38	5.0
13	1	247	25	0.43	5.4
14		222	24		
15	2	198	1	1.01	6.4
16		197	11		
17	3	186	7	1.64	8.0
18	1	179	12	0.58	8.6
19	1	167	14	0.63	9.2
20		153	12		
21		141	14		
22	3	127	11	2.47	11.7
23	1	116	13	0.91	12.6
24	1	103	9	1.02	13.6
25	1	94	8	1.11	14.8
26	1	86	12	1.25	16.0
27		74	13		
28		61	11		
29		50	9		
30		41	13		
31		28	13		
32		15	12		
33		3	3		

Data. Such an estimate is illustrated with the design failure modes of the traction motors (modes A, B, C, D, E, J, M, N, P, Q, R, S, and T). Engineers proposed to eliminate modes A, B, and C through design improvements; the resulting design MCF must be estimated to judge whether eliminating them is worthwhile. Here the resulting group of remaining failure modes is (D, E, J, M, N, P, Q, R, S, and T).

Remaining MCF. The sample MCF for a group G of remaining types of events is calculated as follows. Use all sample censoring ages, but use only the recurrence data on the remaining types of events comprising the group. Combine all data on those remaining recurrences, ignoring all other recurrences. This is shown in Table 6.5 for the design modes with A, B, and C eliminated. The second column contains the total number of remaining design failures observed in each month. Each monthly total is obtained by summing across all remaining design modes in that month's row of Table 1.6. The calculation of the MCF estimate (expressed as a percentage) in Table 6.5 is described in Chapter 5 on interval data. This sample MCF is in Figure 6.3, labeled design without ABC.

Interpretation. At 33 months, the remaining MCF value is 33%. Figure 6.3 also shows the MCF for all design modes and the MCF for modes ABC. At 33 months, the design

6.7. Practical and Theoretical Issues

Table 6.5. *MCF calculation for design modes with A, B, and C eliminated.*

Month i	Number of design minus A, B, C	Number enter month	Number censored	Increment % m_i	MCF % M_i^*
1		372			
2	1	372		0.27	0.3
3	2	372		0.54	0.8
4	4	372		1.08	1.9
5		372			
6	1	372		0.27	2.2
7	1	372		0.27	2.4
8	1	372	13	0.27	2.7
9	2	359	28	0.58	3.3
10	2	331	28	0.63	3.9
11	3	303	29	1.04	4.9
12	1	274	27	0.38	5.3
13	2	247	25	0.85	6.2
14	3	222	24	1.43	7.6
15	1	198	1	0.51	8.1
16	4	197	11	2.09	10.2
17		186	7		
18	2	179	12	1.16	11.4
19	3	167	14	1.88	13.2
20	3	153	12	2.04	15.3
21	1	141	14	0.75	16.0
22	2	127	11	1.65	17.7
23	1	116	13	0.91	18.6
24	4	103	9	4.06	22.6
25	1	94	8	1.11	23.8
26	1	86	12	1.25	25.0
27	2	74	13	2.96	28.0
28	1	61	11	1.80	29.8
29		50	9		
30	1	41	13	2.90	32.7
31		28	13		
32		15	12		
33		3	3		

MCF is 49%. Thus eliminating modes ABC reduces the design MCF to $33/49 = 68\%$ of its original value. Figure 6.3 shows that the recurrence rate (derivative) for the remaining design modes initially increases and then becomes essentially constant, much like the rate for all modes.

Enhancements. Such a plot can be enhanced. As before, one can add confidence limits and display the censoring times. Also, the plotted points can be symbols that identify the remaining types of events, for example, $(D, E, \ldots, T)$. This reveals which types of events dominate as the motors age.

Extensions. As above, this remaining MCF estimate extends to cost (or value) data, exact age data, age data with left censoring and gaps, and continuous history functions. Confidence limits can also be calculated and plotted. Since the data here are interval data, only the naive limits can be calculated. For each type of data, use the corresponding estimate from Chapter 3 and confidence limits from Chapter 4.

6.7 Practical and Theoretical Issues

Purpose. This section discusses practical and theoretical issues that affect the reliability of the methods above.

Assumptions. The MCF estimates and confidence limits of previous chapters are used for data with a mix of events. The assumptions of those methods must be assessed as described in those chapters. The mix of types of recurrences poses some further issues.

Cause and effect. In some applications, one event may cause another event. For example, failure of component A in a product may overload component B and cause it to fail. Then should the two events be treated as a single event or two events? The answer depends on the purpose of the analysis. For purposes of evaluating the reduction in repair costs due to redesign of component A, the costs of associated component B repairs should be included. For purposes of predicting needed replacements for component B, all replacements of component B should be treated as separate events, whether caused by component A or some other cause.

Independence. The different types of events need not be statistically independent. The analyses above do not use this assumption.

Errors. In some data sets, there is a possibility that some events are assigned to the wrong type of event. Such reporting errors occur in many applications and add to the uncertainty of results.

Literature. There is little literature on methods for recurrence data with a mix of events. For example, Ng and Cook (1999) consider robust analysis of a mix of types of events that may be statistically dependent. In contrast, there is much literature on the analogous problem of competing risks for life data; for example, Crowder (2001), Nelson (1982, Chapter 5), Meeker and Escobar (1998, Chapter 15), and David and Moeschberger (1978) present the basic series-system model and data analyses.

Problems

6.1. Traction motors. There were many other analyses of interest, such as the following:

(a) Calculate the naive 95% confidence limits for the MCF with all failure modes. Plot them on Figure 6.1.

(b) Calculate the naive 95% confidence limits for the MCF for failure mode K. Plot them on Figure 6.2.

(c) Choose one of the more frequent failure modes and calculate its sample MCF. Plot that MCF on Figure 6.2 with mode K. Discuss how the two MCFs compare.

(d) Calculate the sample MCF for the combined nondesign modes F, G, H, I, L, O, and X and plot it. (The cause of mode K was unknown and could not be assigned to the design or nondesign group.) Comment on the behavior of their recurrence rate.

(e) Calculate the naive 95% confidence limits for the MCF for nondesign modes and plot them on your MCF plot.

6.2. Mode K. Spread the mode K repairs over their corresponding intervals and treat the repair ages as exact.

(a) Calculate the MCF estimate for the "exact" data, and plot it on Figure 6.2. Comment on how the two plots compare. Which do you prefer and why?

(b) Calculate the naive 95% limits for the exact data, and plot them on Figure 6.2. Comment on how the two sets of limits compare. Which do you prefer and why?

Chapter 7
Comparison of Samples

7.1 Introduction

Purpose. This chapter presents graphical methods for comparing sample MCFs to assess if they differ statistically significantly, that is, convincingly. These methods are illustrated with the data on repairs of automatic and manual transmissions of cars, treatments for recurrent bladder tumors, and two groups of braking grids in locomotives. These methods apply to numbers and costs (or values) of events, and costs may be negative. For example, Morin (2002) compares the purchasing behavior of groups of Internet shopping customers receiving different sales promotions.

Overview. This chapter covers the following topics:

7.1 *Introduction*: This section overviews the chapter and presents confidence limits for the difference of two MCFs.

7.2 *Pointwise comparison of MCFs*: This section first presents methods for comparing two sample MCF values at a specified age (point in time). These methods employ a plot of the pointwise difference of the sample MCFs and corresponding confidence limits. A number of applications illustrate such comparison plots. This section explains the properties of such plots, surveys software for them, and extends comparisons to MCF values of K samples at a specified age.

7.3 *Comparison of entire MCFs*: This section presents methods for comparing MCFs over their entire age range. They employ a statistic that sums a weighted difference of MCFs over their common age ranges.

7.4 *Theory, issues, and extensions*: This section presents the assumptions for such comparisons, practical issues, and extensions of the comparisons to other types of data (e.g., left censored data, data with gaps, interval data, and continuous history functions).

Simple theory. Suppose that samples 1 and 2 yield nonparametric sample MCFs $M_1^*(t)$ and $M_2^*(t)$. To compare them at an age t, we look at their difference $[M_1^*(t) - M_2^*(t)]$ and two-sided $C\%$ confidence limits for the difference. If the limits do not enclose zero, that difference is statistically significant. Otherwise, the two sample MCFs do not differ convincingly at that age. Under the assumptions stated in section 7.4 which are usually satisfied in practice, the sampling distribution of $M_1^*(t) - M_2^*(t)$ is approximately normal with mean $M_1(t) - M_2(t)$ and variance $V[M_1^*(t)] + V[M_2^*(t)]$. Chapter 4 gives Nelson's unbiased estimator $v[M_i^*(t)]$ for $V[M_i^*(t)]$. Thus two-sided approximate $C\%$ confidence limits for $M_1(t) - M_2(t)$ at age t are

$$[M_1^*(t) - M_2^*(t)] \pm z_{C'}\{v[M_1^*(t) - v[M_2^*(t)]\}^{\frac{1}{2}}. \tag{7.1}$$

Here $z_{C'}$ is the $C' = \frac{1}{2}(100 + C)$th standard normal percentile. Note that these are *pointwise* limits. That is, at any one age t, the limits enclose the corresponding population difference $[M_1(t) - M_2(t)]$ with (approximate) probability $C\%$. They do *not simultaneously* enclose the function $M_1(t) - M_2(t)$ at *all* ages in the range of the data with probability $C\%$. Thus the pointwise limits must not enclose zero over some reasonable range of the data to indicate that the MCFs differ convincingly.

Extensions. The limits (7.1) apply to the common situation with exact age data and right censoring and to counts and costs (values) of recurrences. The limits also extend to exact age data censored on the left and with gaps. Other variance estimates referenced in Chapter 4 could be used. The limits (7.1) can also be used for interval age data and counts of recurrences as described in Chapter 5. Then one must use the naive variance estimate.

7.2 Pointwise Comparison of MCFs

Purpose. This section presents pointwise comparisons of two sample MCFs, $M_1^*(t)$ and $M_2^*(t)$, at a single age t. This comparison employs a plot of the difference of the two MCFs and the corresponding confidence limits (7.1). These plots are illustrated with three applications—transmissions, bladder tumors, and braking grids. When there are more than two sample MCFs, they can be compared pairwise as described below.

Transmission Application

Transmission data. The data on manual and automatic transmissions appear in Table 1.1 and Problem 3.1. Their sample MCFs appear in Figures 7.1(a) and (b). Examination of the MCF plots shows that, at any age, the 95% confidence limits for one MCF enclose the estimate for the other MCF. Thus the two MCFs do not have a statistically significant difference at any age. This comparison is easier to make if both MCFs and confidence limits are on the same plot, using two colors for the two data sets. Such a simple comparison works when the difference between the MCFs is either negligible or very large compared to the width of the confidence limits.

Difference. Figure 7.1(c) displays the difference between the sample MCF functions, and 95% confidence limits for the difference. For most ages, the limits enclose zero. Thus, this plot also shows that the sample MCFs do not differ significantly over most of the range of the data. Of course, the two samples contain few cars and few repairs, and consequently the confidence limits are wide and the comparison is insensitive. If there is no significant difference, one might pool the two data sets to estimate a common MCF with greater accuracy, if that would be consistent with engineering knowledge.

Plot properties. Such a plot of the difference of two sample MCFs and the corresponding confidence limits has the following properties:

- This nonparametric estimate of the difference is a staircase function that has a jump at each recurrence in each sample. The difference is a horizontal line segment extending to the right of a recurrence up to the next recurrence. Often these segments are not plotted, as they obscure the data points.

7.2. Pointwise Comparison of MCFs   111

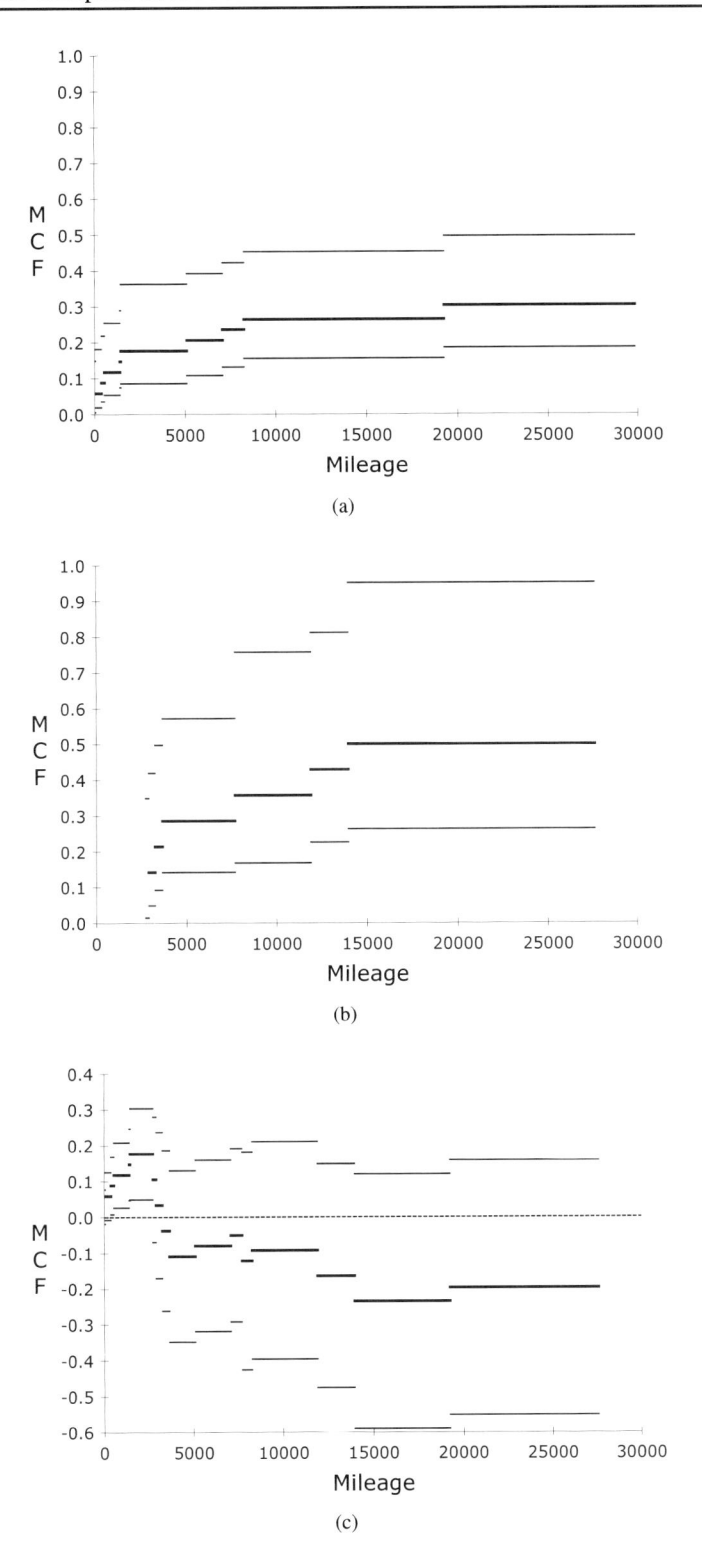

Figure 7.1. (a) *Automatic transmission MCF and 95% limits.* (b) *Manual transmission MCF and 95% limits.* (c) *MCF difference (automatic–manual) and 95% limits.*

- Similarly, the confidence limits are staircase functions that jump at each recurrence in each sample. They, too, extend horizontally to the right of a recurrence up to the next recurrence. Often these segments are not plotted.

- Like confidence intervals for a single MCF, these intervals for the difference get larger as t increases, for two reasons. First, the true standard deviation of the difference typically increases as t increases. Second, the number of sample units reaching higher ages is small, resulting in wider intervals.

Such a plot is preferable to a hypothesis test, whose test statistic merely says whether the difference is convincing or not. The more important question of how the MCFs differ can be best answered by examination of the plot. Plots and confidence limits used for a comparison are always more informative than a hypothesis test.

Software. The following programs calculate and plot the difference and its confidence limits from exact age data with right censoring. They provide numerical output as well as plots. They also apply to cost (or value) data on recurrences and allow for negative values.

- MCFDIFF of Doganaksoy and Nelson (1991).

- The Reliability Procedure in the SAS/QC software of the SAS Institute (1999, pp. 947–951). This provided Figures 7.2(a)–(c) below.

- In 2002, such features will be available in the JMP software of the SAS Institute. JMP provided Figures 7.3(a)–(c) below.

- SPLIDA features developed by Meeker and Escobar (2002) for the S-PLUS software of Insightful (2001).

- The recurrence data add-on of the ReliaSoft Corporation (2000a, b) Weibull++ software. This provided Figures 7.1(a)–(c).

Bladder Tumor Application

Bladder tumor data. The data on bladder tumor recurrences for placebo and Thiotepa treatments appear in Table 1.2 and Problem 3.2. Their sample MCFs appear in Figures 7.2(a) and (b). Figures 7.2(a)–(c) were produced by the SAS code in Figure 4.3(c). Examination of the linear MCF plots and 95% confidence limits does not clearly indicate whether the two distributions differ statistically significantly at any age. (The comparison is easier to make if both MCFs appear on the same plot in different colors.) Thus one must look at their difference. The slopes of the plots seem to be quite different. An objective comparison of the slopes appears in Problem 8.2.

Treatment difference. Figure 7.2(c) displays the difference between the sample MCF functions and 95% confidence limits for the difference. At every age, the limits enclose zero. Thus this plot also shows that the sample MCFs do not significantly differ pointwise over the range of the data. Problem 8.2 provides another comparison of the two treatments.

Braking Grid Application

Braking grids. JMP of the SAS Institute (2000) provided Figures 7.3(a) and (b). They show sample MCFs for locomotive braking grid replacements on production groups A and B. The data are from Doganaksoy and Nelson (1998). Group B accumulates replacements linearly starting at age 0 days. Group A accumulates replacements linearly with the same rate

7.2. Pointwise Comparison of MCFs

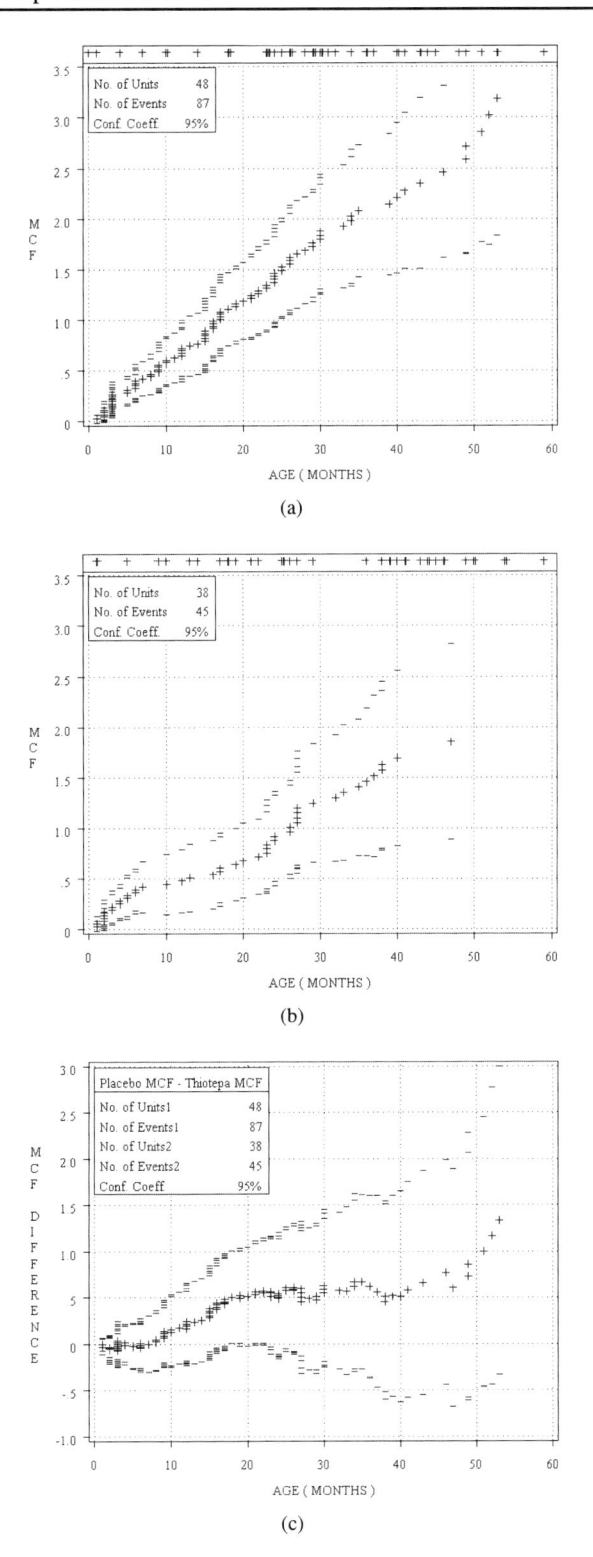

Figure 7.2. (a) *Placebo MCF and 95% confidence limits.* (b) *Thiotepa MCF and 95% confidence limits.* (c) *MCF difference (placebo–Thiotepa) and 95% limits.*

114   Chapter 7. Comparison of Samples

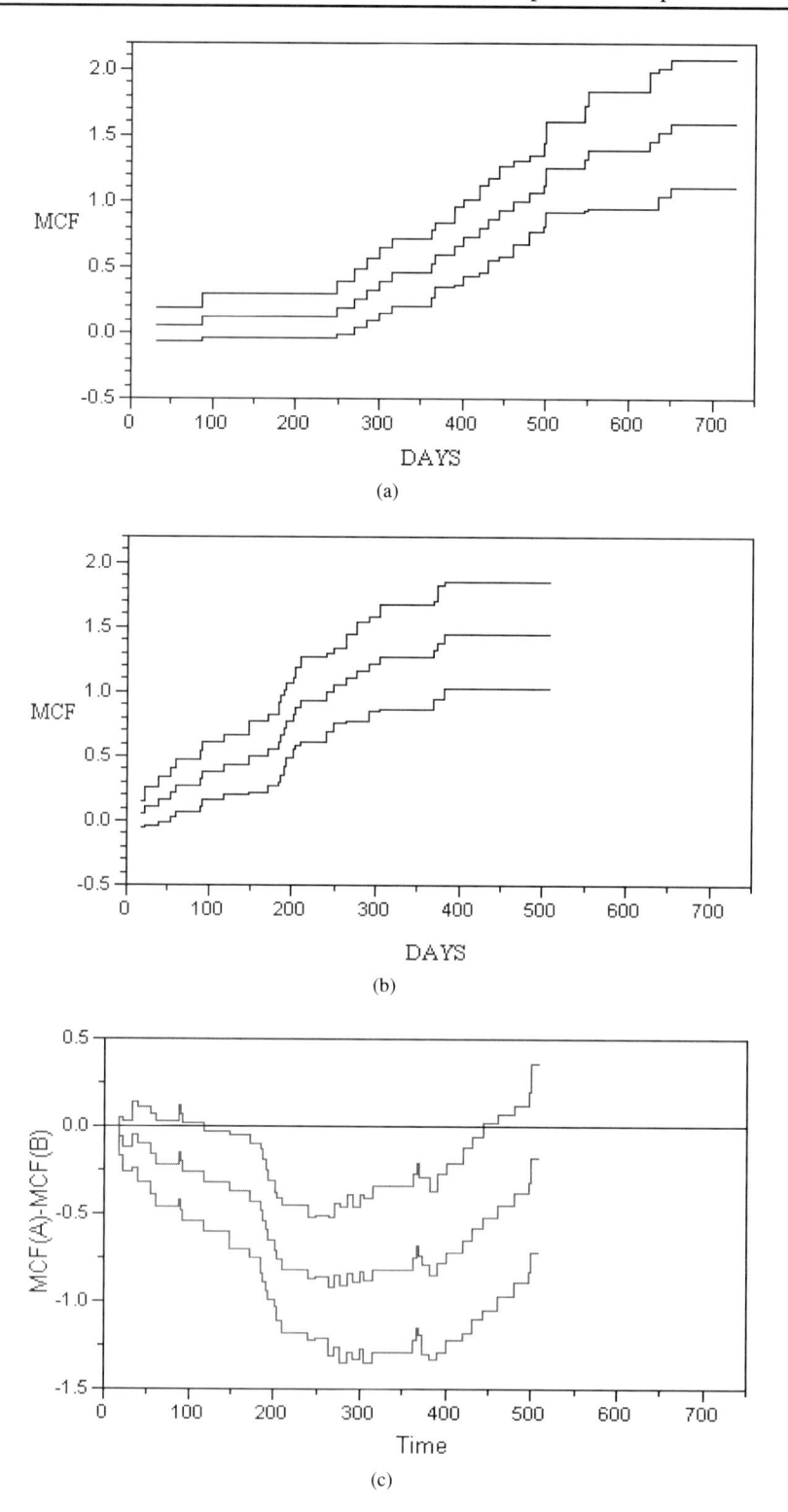

Figure 7.3. (a) *MCF and 95% limits of batch* A *braking grids.* (b) *MCF and 95% limits of batch* B *braking grids.* (c) *Groups MCF difference* (A–B) *and 95% limits.*

7.2. Pointwise Comparison of MCFs

(slope) but starting at 250 days. Both groups experienced the same operating conditions, and each locomotive has six grids. Before investigating the cause of the difference, the responsible engineers wanted to know whether the observed difference was convincing, that is, statistically significant. The confidence limits of the MCF plots appear not to overlap; this suggests a convincing difference.

Group difference. Figure 7.3(c) is a JMP plot of the MCF difference and is more revealing. Clearly, the confidence limits do not enclose 0 over most ages. Thus convinced of a real difference, the engineers investigated. They found that the supplier had used a different stamping die to cut the group B grids, and the new die did not meet tolerances. They replaced it and eliminated such early failures.

Permutation tests. The preceding pointwise comparison employs approximate confidence limits. A more exact pointwise comparison at age t is achieved with a permutation test, as follows. Suppose that the two samples have N_1 and N_2 units. Pool the two samples. To get sample 1, take N_1 units from the $N_1 + N_2$ units; sample 2 is the remaining N_2 units. Then calculate the observed difference $M_1^*(t) - M_2^*(t)$ for that pair of samples. Repeat this to get a difference for every pair of such samples. If the actual observed difference is in a far tail of this sampling distribution of these differences, then it is statistically significant. In practice, the number $(N_1 + N_2)!/N_1!N_2!$ of such samples may be too large to enumerate. Then one can approximate the sampling distribution with simulation. This involves repeatedly taking a random sample of N_1 units from the $N_1 + N_2$ units and calculating the $M_1^*(t) - M_2^*(t)$ of the two samples. Such a permutation test is most natural when sample units are randomly assigned to treatment groups, as were the patients in the bladder tumor study. Instead of the simple difference, one can use some other reasonable statistic, such as the difference divided by an estimate of its standard error. Software for such permutation tests has not been developed.

All pairwise differences. When there are K sample MCFs $M_k^*(t)$, $k = 1, 2, \ldots, K$, examine the $K(K-1)/2$ plots of the difference of each pair of MCFs, each with its corresponding $C\%$ confidence limits, to see which enclose zero. That is, examine all intervals

$$[M_k^*(t) - M_{k'}^*(t)] \pm z_{C'}\{v[M_k^*(t)] + v[M_{k'}^*(t)]\},$$

where $z_{C'}$ is the $C' = \frac{1}{2}(100 + C)$th standard normal percentile. For example, Agrawal and Doganaksoy (2001) compare $K = 3$ sample MCFs in this way. If all true differences were zero, the probability that some interval would fail to enclose zero would exceed $(100 - C)\%$, because there is more than one interval.

Simultaneous intervals. We can use wider simultaneous intervals, which have approximate probability $C\%$ that *all* $K(K-1)/2$ intervals simultaneously enclose their corresponding true differences. Under this more stringent criterion, if a simultaneous interval does not enclose zero, we have stronger evidence of a real difference. Then we say that there is a *wholly significant difference*. Such intervals are obtained by replacing C' above with $C'' = 100 - 2(100 - C)/[K(K-1)]$. These simultaneous limits employ the Bonferroni inequality, a conservative approximation that yields intervals that *all simultaneously* enclose their true differences with confidence at least $C\%$. It is useful to examine a plot of all such differences with their corresponding single and simultaneous confidence limits.

Analysis-of-variance comparison. The following analysis-of-variance test can be used to simultaneously compare K independent sample MCF values $M_1^*, M_2^*, \ldots, M_K^*$. Suppose that the corresponding estimates of their variances are $v_1, v_2, \ldots, v_K$. Calculate the pooled estimate of a common MCF value

$$M^* = v \cdot \left[\left(\frac{M_1^*}{v_1}\right) + \left(\frac{M_2^*}{v_2}\right) + \cdots + \left(\frac{M_K^*}{v_K}\right)\right],$$

where

$$v = \frac{1}{\left[\left(\frac{1}{v_1}\right) + \left(\frac{1}{v_2}\right) + \cdots + \left(\frac{1}{v_K}\right)\right]}$$

is an estimate of the variance of M^*. Then calculate the quadratic statistic

$$Q = \left[\frac{(M_1^* - M^*)^2}{v_1}\right] + \left[\frac{(M_2^* - M^*)^2}{v_2}\right] + \cdots + \left[\frac{(M_K^* - M^*)^2}{v_K}\right].$$

When the M_k^* are approximately normally distributed, this statistic is approximately chi-square distributed with $K - 1$ degrees of freedom. Then the approximate test for equality of the K population MCF values is as follows:

- If $Q \leq \chi^2(100 - \alpha; K - 1)$, the K estimates do not differ statistically significantly (convincingly) at the $\alpha\%$ level.

- If $Q > \chi^2(100 - \alpha; K - 1)$, they differ statistically significantly at the $100\alpha\%$ level.

Here $\chi^2(100 - \alpha; K - 1)$ is the $(100 - \alpha)$th chi-square percentile with $K - 1$ degrees of freedom. If the estimates differ significantly, then it is necessary to examine them and their confidence limits to determine how they differ. This can be done with a plot of all estimates and their limits. Similarly, one can examine all pairwise differences and corresponding limits to determine how the estimates differ.

Parametric comparisons. All comparisons in the first seven chapters were nonparametric. Parametric models can be used to compare data sets. Such models are fitted by maximum likelihood to each data set, as described by Rigdon and Basu (2000). The fitted model for each data set yields a parametric estimate of its sample MCF at a specified age, and these estimates can be compared with likelihood ratio tests. When suitable, parametric methods make full use of all data over the entire age range of each sample and yield more accurate estimates. In contrast, the nonparametric comparisons here use the sample data only up to the specified age.

7.3 Comparison of Entire MCFs

Purpose. The previous section deals with comparing two or more sample MCFs *at a single age*. This section discusses comparison of sample MCFs *over their entire common age ranges*. A simple visual comparison of plots of MCFs over their age ranges often yields clear conclusions. When such plots are not conclusive, suitable formal numerical comparisons are needed. Cook, Lawless, and Nadeau (1996) and the references therein present such comparisons under various assumptions. This section merely discusses comparisons in general terms.

Plots. One can visually examine plots of the individual sample MCFs and their difference with corresponding pointwise confidence limits. This may provide subjective but convincing evidence of a (or no) difference. Figures 7.3(a)–(c) (braking grids) clearly display such a convincing difference. On the other hand, Figures 7.1(a)–(c) (transmissions) clearly show no evidence of a convincing difference. Figures 7.2(a)–(c) (bladder tumors) are inconclusive, and a formal numerical comparison is needed. However, even if a formal statistic shows a convincing difference among sample MCFs, it is still necessary to examine the MCF plots to see what the statistic is detecting.

7.3. Comparison of Entire MCFs

Overview. This section first considers the simple comparison of two sample MCFs and then considers K MCFs. Such comparisons are analogous to those for comparing two or K multivariate means. The section describes a general approach to such a comparison. Then it proposes some statistics for such comparisons. Details for such comparisons have not yet been worked out.

Approach. First, one must have a suitable statistic that is a measure of the difference of (or distance between) two sample MCFs across the overlap of their age ranges. Data outside the overlap cannot be used in a nonparametric analysis. Next, one must obtain the sampling distribution of the statistic through an approximation, permutation methods, or a simulation method like bootstrapping. If the observed value of the statistic is in a far tail of the sampling distribution, it indicates a statistically significant (convincing) difference of the two sample MCFs. To determine what the statistic has detected, it is then necessary to examine plots of the individual MCFs and their difference, with corresponding confidence limits.

Two-sample statistic. A comparison of two sample MCFs is analogous to a multivariate comparison of two mean vectors, using a statistic like Hoetelling's T^2. The following discussion is an outline of work that needs to be done, including software. For simplicity, suppose that two samples are statistically independent and have distinct exact age data with right censoring. (We ignore age ties here.) Denote their MCF estimates by $M^*(t)$ and $N^*(t)$. Suppose that the sample with the shorter age range has $I + 1$ observed recurrence ages $t_1, t_2, \ldots, t_{I+1}$. We hereafter ignore t_{I+1} since we cannot estimate the variance of $M^*(t_{I+1})$. Suppose that the other sample has J recurrences in the common age range at ages $s_1, s_2, \ldots, s_J$. For simplicity, assume that s_J is not the last recurrence age for that sample. Let D_d denote the $I + J$ MCF differences at ages $t_1, t_2, \ldots, t_I, s_1, s_2, \ldots, s_J$. Here the index d runs over the $I + J$ ordered ages. For the first sample, let V denote an estimate of the covariance matrix of the I MCF estimates $M^*(t_1), \ldots, M^*(t_I)$. For the second sample, let W denote an estimate of the covariance matrix of the J MCF estimates $N^*(s_1), \ldots, N^*(s_J)$. These two covariance matrices can be used to calculate an estimate of the covariance matrix U of the row vector D of d differences. Then the quadratic statistic to test whether the differences are all zero is

$$Q = DU^{-1}D',$$

where prime (′) denotes transpose. This statistic is analogous to Hoetelling's T^2 statistic for comparing mean vectors. The sampling distribution of Q must be approximated and used as described above.

Other statistics. Other statistics could be used. For example, to simplify, one might assume that the processes have independent increments. Then the covariance matrices V and W are diagonal, and the calculation of Q is simpler.

K samples. The preceding approach extends to comparison of K samples. Then one must calculate a row vector D of all differences of each pair of MCF estimates. Using the K covariance matrix estimates $V_1, V_2, \ldots, V_K$ of the K sample MCFs, one can calculate an estimate of the covariance matrix U of D. Then one uses a Q statistic like that above to test for equality.

Parametric comparison. Parametric comparison of sample MCFs involves testing whether corresponding parameter estimates for the sample MCFs differ significantly. Such comparisons use all data from each sample, not just from the overlap of the age ranges. Rigdon and Basu (2000) present such parametric comparisons using likelihood ratio tests.

7.4 Theory, Issues, and Extensions

Purpose. This section briefly reviews the assumptions underlying the limits (7.1), extensions to other types of data, and confidence bands.

Assumptions. The nonparametric confidence limits (7.1) for the difference of two sample MCFs depend on a number of assumptions. The MCF estimates depend on assumptions ⟨1⟩–⟨6⟩ of Chapter 3, which are summarized in section 4.4. The variance estimates depend on assumptions ⟨7⟩–⟨9⟩ of section 4.4. Similar assumptions for interval age data are given in Chapter 5. In addition, the limits (7.1) also depend on the following assumptions:

⟨7'⟩ The cumulative sampling distribution of the observed difference $[M_1^*(t) - M_2^*(t)]$ is close to normal. This assumption replaces ⟨7⟩; each estimator $M_i^*(t)$ has a sampling distribution that is close to normal. In practice, $M_1^*(t)$ and $M_2^*(t)$ usually have similar sampling distributions when their sample sizes, data, and censoring are similar. Moreover, when the two sampling distributions are skewed, they are usually skewed the same way. Then their difference has a sampling distribution that is closer to symmetric and closer to normal than the individual sampling distributions. Thus ⟨7'⟩ is better satisfied for smaller samples and for more situations than is ⟨7⟩.

⟨10⟩ The two samples are statistically independent. This is used to obtain the variance result $V[M_1^*(t) - M_2^*(t)] = V[M_1^*(t)] + V[M_2^*(t)]$, used in the limits (7.1).

Data. The limits (7.1) apply to the common situation with exact age data and right censoring and to counts and costs (values) of recurrences. The limits also extend to exact age data censored on the left and with gaps. Other variance estimates referenced in Chapter 4 could be used. The limits (7.1) can also be used for interval age data and counts of recurrences as described in Chapter 5; then one must use the naive variance estimate.

Confidence bands. The confidence limits (7.1) enclose the difference of two MCFs at a *single* age with probability $C\%$. In contrast, confidence bands simultaneously enclose all differences over the common range of the two MCFs with probability $C\%$. Consequently, they are wider. It may be possible to extend Vallarino's (1988) confidence bands for a single MCF to the difference of two. With such bands, an observed difference is more convincing. His bands depend on the assumption of a nonhomogeneous Poisson process and the naive variance estimate.

Problems

7.1. Childbirths. Use the men's and women's childbirth data in Table 1.5.

(a) Calculate and plot the difference of the men's and women's MCFs.

(b) Calculate and plot the naive approximate 95% confidence limits for the difference. Comment on whether and how the MCFs differ statistically significantly.

7.2. Chronic granulomatous disease. Fleming and Harrington (1991, pp. 162–163 and 376–383) give data on recurrences of serious infections of patients with CGD. Patients were randomly assigned to gamma interferon treatment and to placebo treatment.

(a) Calculate and plot the MCF and 95% confidence limits for each treatment group,

Problems

ignoring covariates. Compare the two MCFs. Do you think that the difference is convincing? Why?

(b) Calculate and plot the difference of the treatment MCFs with 95% confidence limits. In view of this plot, do you think that the difference is convincing? Why?

(c) Use the covariates to divide the data into covariate groups, for example, using or not using corticosteroids, using or not using antibiotics, male or female, etc. Calculate and plot the MCF for each group. Note the differences that you think are convincing and why.

(d) For the covariate groups, calculate and plot the difference of the treatment MCFs with 95% confidence limits. In view of these plots, which differences do you think are convincing? Why?

7.3. Naive comparison. Develop an approximate hypothesis test to compare two entire sample MCFs where they overlap as follows. Assume that the two underlying populations are nonhomogeneous Poisson processes.

(a) Suppose that the data are interval data with the same set of age intervals for the two samples. Devise a test statistic and determine its approximate sampling distribution under the null hypothesis of equal population MCFs. State all assumptions.

(b) Apply your test statistic to comparing the childbirth MCFs for men and women. State your conclusions.

(c) Suppose that the data are exact age data. Devise a test statistic and determine its approximate sampling distribution under the null hypothesis of equal population MCFs. State all assumptions.

(d) Apply your test statistic to comparing the manual and automatic transmission MCFs. State your conclusions.

Chapter 8
Survey of Related Topics

8.1 Introduction

Purpose. This chapter is a brief introduction to related topics, in particular, basic parametric models for the *number* of recurrences. These counting process models require more assumptions that the nonparametric model above. Moreover, they apply only to the *number* and not costs or other measures of recurrences.

Choice of model. In practice, one must decide when to use the nonparametric model above and when to use a parametric model. Indeed, one should always make the MCF plot, as it reveals information that parametric model-fitting does not. When a parametric model has been assessed and found suitable, it will give more accurate estimates. More important, it gives added insight or a simple, useful way of thinking about the data.

Overview. The contents of this chapter are as follows:

8.2 *Poisson process*: This simplest parametric model for the number of recurrences has a linear MCF $M(t) = \lambda t$ and a constant recurrence rate $m(t) = \lambda$.

8.3 *Nonhomogeneous Poisson process*: This model is a generalization of the Poisson process with an arbitrary MCF $M(t)$.

8.4 *Renewal processes*: When a failed unit is immediately replaced with a new one, the population $M(t)$ depends on the cumulative life distribution $F(t)$ of the units.

8.5 *Models with covariates*: In many applications, the recurrence rate is expressed as a function of covariates. Examples here include the Poisson regression model and the Cox model for recurrences.

8.6 *Other models*: This briefly surveys other models for recurrence data.

Problems.

8.2 Poisson Process

Purpose. This section briefly presents the Poisson process model and data analyses. It is the simplest counting process model for the number Y of recurrences of an event in an observed time t (or area, volume, etc.). For example, it has been used to model the yearly number of soldiers of a Prussian regiment kicked to death by horses, the number of flaws in a length of wire or computer tape, the number of defects in a sheet of material, the number of repairs of a product over a certain period, the number of atomic particles emitted by a sample in a specified time, and many other counting phenomena. Although it is not suitable for any data in this book, it is useful for constructing more realistic models. For more details on the Poisson process and data analyses, see Nelson (1982).

Overview. This section covers the following topics:

8.2.1 *Poisson process model*: This defines and gives the properties of this most basic model.

8.2.2 *Single-sample analyses*: This gives Poisson estimates and confidence limits and predictions and prediction limits.

8.2.3 *Multisample data analyses*: This gives Poisson analyses for pooling and comparing samples.

8.2.1 Poisson Process Model

Introduction. This subsection presents the Poisson process model and its properties. It is a model for a single unit or a population of units all with the same constant recurrence rate λ, which is rare in practice. The topics below include a definition of the process, the Poisson density and cumulative distribution, the mean, variance and standard deviation, interarrival times and the exponential distribution, and sums of Poisson counts.

Poisson process. In some applications, recurrences are observed at random points in time. The Poisson process model describes many such situations. These include, for example, (1) failures in a stable fleet of repairable items, (2) phone calls coming into an exchange, (3) atomic particles registering on a counter, and (4) power line failures. In some books, the Poisson process is presented as a birth process.

Definition. The model is defined by the following properties: (1) the potential number Y of recurrences in any time period of length t is unlimited, (2) the numbers of recurrences in any number of nonoverlapping intervals are all statistically independent ("independent increments"), and (3) the chance of a recurrence is the same for each point in time. (Parzen (1999) gives a more formal definition.) Consequently the number Y of recurrences in any period of length t has a Poisson distribution with mean $\mu = \lambda t$, where the positive parameter λ is the recurrence rate.

Density. The Poisson density function for the probability of y recurrences is

$$f(y) = \left(\frac{1}{y!}\right)(\lambda t)^y \exp(-\lambda t), \quad y = 0, 1, 2, \ldots. \tag{8.1}$$

The dimensions of λ are recurrences per unit time per population unit. For example, for the Proschan data (Problem 2.2), the dimensions of λ are repairs per hour per air conditioner. Many authors present the Poisson distribution with a single parameter $\mu = \lambda t$. Figure 8.1 depicts Poisson probability densities for various μ values. Figure 8.2 depicts the Poisson process with $M(t) = \lambda t$ and shows the Poisson distribution of the cumulative number of recurrences at two ages.

Mean. The Poisson mean number Y of recurrences over a time t is

$$E(Y) = \lambda t. \tag{8.2}$$

8.2. Poisson Process

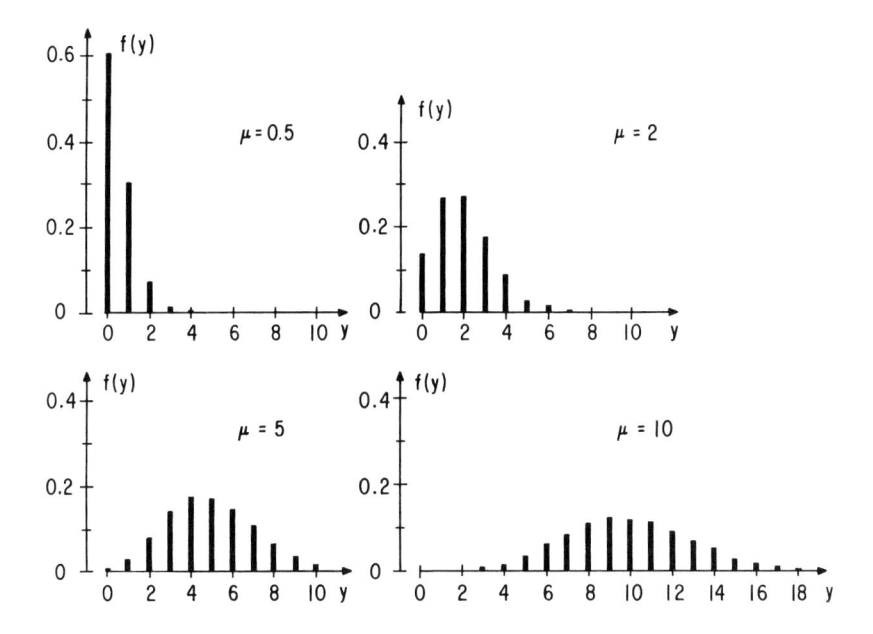

Figure 8.1. *Poisson probability densities.*

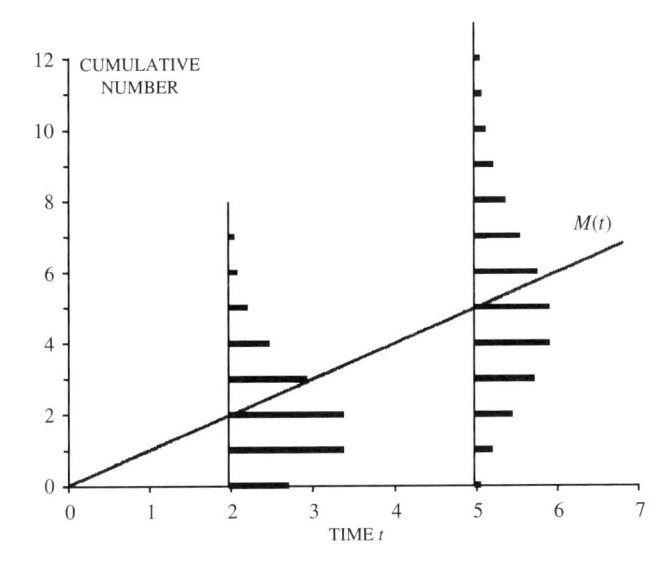

Figure 8.2. *Poisson process depicted.*

Thus the Poisson process MCF is $M(t) = \lambda t$. Rewriting (8.2) yields

$$\lambda = \frac{E(Y)}{t}. \tag{8.3}$$

This shows why λ is called the recurrence rate; it is the expected number of recurrences divided by the corresponding time t.

Moments. The variance and standard deviation of the number Y of recurrences are

$$\text{Var}(Y) = \lambda t, \qquad \sigma(Y) = (\lambda t)^{\frac{1}{2}}. \tag{8.4}$$

Note that the Poisson variance and mean both equal λt.

Cdf. The cdf (the probability of y or fewer recurrences) is

$$F(y) \equiv \Pr\{Y \le y\} = \sum_{i=0}^{y} \left(\frac{1}{i!}\right) (\lambda t)^i \exp(-\lambda t). \tag{8.5}$$

A normal approximation to this is

$$F(y) \cong \Phi\left\{\frac{[y + 0.5 - EY]}{\sigma(Y)}\right\} = \Phi\left[\frac{y + 0.5 - \lambda t}{(\lambda t)^{\frac{1}{2}}}\right]; \tag{8.6}$$

here $\Phi(\cdot)$ is the standard normal cdf and is tabulated in many statistics textbooks. This approximation is the more exact, the larger λt is and the closer y is to λt. It is satisfactory for many practical purposes if $\lambda t \ge 10$.

Times between recurrences. Some recurrence models are described in terms of times D_i between recurrences, called interarrival times. For a Poisson process with recurrence rate λ, these times are from the same exponential distribution with rate λ and are statistically independent. That is, $\Pr\{D_i \le t\} = 1 - \exp(-\lambda t)$, $i = 1, 2, 3, \ldots$. In reliability work, the mean of this distribution $\theta = 1/\lambda$ is called the *mean time between failures* (MTBF).

Sums of counts. The sum of counts of *independent* Poisson counts has a Poisson distribution. In particular, suppose that count k is Y_k, has recurrence rate λ_k, and has observation time t_k, $k = 1, \ldots, K$. Then the sum $Y = Y_1 + \cdots + Y_K$ has a Poisson distribution with mean $\mu = \lambda_1 t_1 + \cdots + \lambda_K t_K$. For example, the number of repairs of the fleet of air conditioners has a Poisson distribution. For the special case $\lambda_1 = \lambda_2 = \cdots = \lambda_K = \lambda$, a common rate, Y has a Poisson distribution with mean $\mu = \lambda t$, where $t = t_1 + t_2 + \cdots + t_K$. This result applies to a single unit observed over K periods $(t_1, t_2, \ldots, t_K)$ or to K units.

8.2.2 Single-Sample Analyses

Purpose. This section gives an estimate and confidence intervals for the Poisson recurrence rate λ. Also, it gives predictions and prediction limits for the number X of recurrences in a future sample.

Data. Poisson data for a single unit consist of the number Y of recurrences in an observed time t. Such data typically come from a single unit, for example, from a particular tumor patient or air conditioner. For example, plane 7907 had $Y = 6$ repairs in $t = 493$ hours (Proschan's data of Problem 2.2). Proschan (2000) modeled air conditioner repairs in each plane with a Poisson process with a different λ_k. He also failure censored his data, whereas the Poisson data analysis methods below are suited to time censored data. This detail is ignored hereafter.

Estimate λ. The estimate for the true λ is the sample recurrence rate

$$\lambda^* = \frac{Y}{t}. \tag{8.7}$$

For example, for plane 7907, $\lambda^*_{7907} = 6/493 = 0.012$ or 1.2 repairs per hundred hours. Also, for all 13 planes, the total number of repairs is $Y = 213$, and the total observed hours is

8.2. Poisson Process

$t = 19839$. The pooled estimate of a common rate is $\lambda^* = 213/19839 = 0.0107$ or 1.07 repairs per plane per hundred hours. λ^* is unbiased ($E\lambda^* = \lambda$), and its variance is

$$\text{Var}(\lambda^*) = \frac{\lambda}{t}. \tag{8.8}$$

The sampling distribution of $Y = \lambda^* t$ is Poisson with mean λt. When the expected number of recurrences λt is large (say, greater than 10), the distribution of λ^* is approximately normal with a mean of λ and variance (8.8).

Confidence limits. Two-sided $100\gamma\%$ confidence limits for λ are

$$\underline{\lambda} = 0.5 \frac{\chi^2\left[\frac{1-\gamma}{2}; 2Y\right]}{t}, \qquad \tilde{\lambda} = 0.5 \frac{\chi^2\left[\frac{1+\gamma}{2}; 2Y+2\right]}{t}; \tag{8.9}$$

here $\chi^2(\delta; v)$ is the 100δth percentile of the chi-square distribution with v degrees of freedom. These limits are conservative; that is, the confidence level is at least $100\gamma\%$. For plane 7907, 95% confidence limits are

$$\underline{\lambda} = 0.5 \frac{\chi^2\left[\frac{1-0.95}{2}; 2\times 6\right]}{493} = 0.5 \frac{\chi^2(0.025; 12)}{493} = \frac{0.5 \times 4.404}{493} = 0.004,$$

$$\tilde{\lambda} = 0.5 \frac{\chi^2\left[\frac{1+0.95}{2}; 2\times 6 + 2\right]}{493} = 0.5 \frac{\chi^2(0.975; 14)}{493} = \frac{0.5 \times 26.12}{493} = 0.0026,$$

or 0.4 and 2.6 repairs per hundred hours. When Y is large (say, $Y > 10$), two-sided approximate $100\gamma\%$ confidence limits for λ are

$$\underline{\lambda} \cong \lambda^* - K_\gamma \left(\frac{\lambda^*}{t}\right)^{\frac{1}{2}}, \qquad \tilde{\lambda} \cong \lambda^* + K_\gamma \left(\frac{\lambda^*}{t}\right)^{\frac{1}{2}}; \tag{8.10}$$

here K_γ is the $100(1+\gamma)/2$th standard normal percentile. Such 95% limits for a common fleet failure rate are

$$\underline{\lambda} \cong 0.0107 - 1.960 \left(\frac{0.0107}{19839}\right)^{\frac{1}{2}} = 0.0093, \qquad \tilde{\lambda} \cong 0.0107 + 0.0014 = 0.0121,$$

or 0.93 and 1.21 repairs per hundred hours.

Prediction. Sometimes one needs to predict a future random number X of recurrences in a future time period of length s. The previous data are Y recurrences in a time t. Since $EX = \lambda s$ and the observed $\lambda^* = Y/t$, the prediction of X is

$$X^* = \lambda^* s = \frac{Y}{t} s. \tag{8.11}$$

X^* is unbiased; that is, $E(X^* - X) = 0$. The prediction error $(X^* - X)$ has variance

$$\text{Var}(X^* - X) = \frac{\lambda s(t+s)}{t}. \tag{8.12}$$

If λt and λs are large (say, both exceed 10), the distribution of $(X^* - X)$ is approximately normal with mean 0 and this variance. For plane 7907 over the next $s = 50$ hours, the prediction is $X^* = (6/493)50 = 0.61$, which is rounded to one repair. However, when predicting the total number of repairs for a fleet of units, the individual decimal predictions are summed and not rounded to integers. For a fleet, the pooled estimate $\lambda^* = 0.0107$ yields a prediction over a future 1000 hours of $X^* = 0.0107 \times 1000 = 10.7$, rounded to 11 repairs, information useful for maintenance planning.

Prediction limits. For two-sided $100\gamma\%$ prediction limits for the future number X of recurrences, the lower limit is the largest integer X_L that satisfies

$$\frac{X_L}{F[\delta; 2Y+2, 2X_L]} \leq (Y+1)\left(\frac{s}{t}\right). \tag{8.13}$$

The upper limit is the smallest integer X_U that satisfies

$$Y\left(\frac{s}{t}\right) \leq (X_U+1) F[\delta; 2X_U+2; 2Y]. \tag{8.13'}$$

Here $F(\delta; a, b)$ is the upper $\delta = 50(1-\gamma)$ percent point of the F distribution with a degrees of freedom in the numerator and b in the denominator. Nelson (1970) gives simple charts for X_L and X_U. To enclose X with probability $100\gamma\%$, the limits must enclose at least $100\gamma\%$ of its distribution. Thus they are much wider than confidence limits (8.9) for its mean $EX = \lambda s$, which is a constant. For plane 7907, the two-sided 90% prediction limits for the future number of repairs over the next $s = 50$ hours are $X_L = 0$ and $X_U = 2$.

Approximate prediction limits. When Y and X^* are large (say, each over 10), two-sided approximate $100\gamma\%$ prediction limits for X are

$$\underline{X} = X^* - K_\gamma \left[\frac{X^*(t+s)}{t}\right]^{\frac{1}{2}}, \quad \bar{X} = X^* + K_\gamma \left[\frac{X^*(t+s)}{t}\right]^{\frac{1}{2}}, \tag{8.14}$$

where K_γ is the $100(1+\gamma)/2$th standard normal percentile. For the fleet over a future 1000 hours, these 95% limits are

$$\underline{X} = 11.6 - 1.960\left[\frac{11.6(19839+1000)}{19839}\right]^{\frac{1}{2}} = 8.1, \quad \bar{X} = 11.6 + 3.5 = 15.1.$$

These are rounded to 8 and 15, information that aids maintenance planning.

8.2.3 Multisample Analyses

Purpose. This subsection provides Poisson analyses for a number of samples of count data. This includes a pooled estimate of a presumed common λ and a comparison of the samples to assess whether they have a common λ.

Data. For $k = 1, \ldots, K$, suppose that Y_k is a Poisson count in an observed time t_k, where the true recurrence rate is λ_k. These may be the data on K units or K samples of units. Also, suppose $Y_1, \ldots, Y_K$ are statistically independent. The following analyses apply to such data. One can combine such Poisson data to get more accurate estimates and predictions as follows.

Pooled λ estimate. To estimate a common Poisson recurrence rate $\lambda = \lambda_1 = \lambda_2 = \cdots = \lambda_K$, first calculate the total number of recurrences $Y = Y_1 + \cdots + Y_K$ and the total observed time $t = t_1 + \cdots + t_K$. Then the pooled estimate of the common λ is

$$\lambda^* = \frac{Y}{t} = \frac{Y_1 + \cdots + Y_K}{t_1 + \cdots + t_K}. \tag{8.15}$$

8.3. Nonhomogeneous Poisson Processes

For example, the 13 planes in Proschan's data had $Y = 213$ repairs in $t = 19839$ hours, and the pooled estimate is $\lambda^* = 213/19839 = 0.0107$ or 1.07 repairs per hundred hours. λ^* is unbiased, and its variance is

$$\text{Var}(\lambda^*) = \frac{\lambda}{t}. \tag{8.16}$$

From (8.10), approximate 95% confidence limits for λ are

$$\underline{\lambda} = 0.0107 - 1.960\left(\frac{0.0107}{19839}\right)^{\frac{1}{2}} = 0.0093, \quad \bar{\lambda} = 0.0107 + 0.0014 = 0.0121,$$

or 0.93 and 1.21 repairs per hundred hours. Before using the pooled λ^*, check whether the K recurrence rates differ significantly, as follows.

Comparison. The following explains how to compare K Poisson recurrence rates for equality. The following tests the equality hypothesis $\lambda_1 = \lambda_2 = \cdots = \lambda_K$ against the alternative $\lambda_k \neq \lambda_{k'}$ for some k and k'. First, calculate the pooled estimate $\lambda^* = Y/t$ for a common λ and the separate $\lambda_k^* = Y_k/t_k$. The test statistic is

$$Q = \sum_{k=1}^{K} \frac{(Y_k - \lambda^* t_k)^2}{\lambda^* t_k} = \sum_{k=1}^{K} (\lambda_k^* - \lambda^*)^2 \frac{t_k}{\lambda^*}. \tag{8.17}$$

If the equality hypothesis is true, the distribution of Q is approximately chi-square with $K - 1$ degrees of freedom. If the alternative is true, Q tends to have larger values. Thus the test is as follows:

- If $Q \leq \chi^2(1 - \alpha; K - 1)$, accept equality.
- If $Q \geq \chi^2(1 - \alpha; K - 1)$, reject equality at the $100\alpha\%$ significance level.

Here $\chi^2(1 - \alpha; K - 1)$ is the $100(1 - \alpha)$th chi-square percentile with $K - 1$ degrees of freedom. Q is called the chi-square statistic. The chi-square approximation is more precise, the larger the Y_k are. It is usually satisfactory if all $Y_k \geq 5$. If there is a statistically significant difference, examine a plot of all the λ_k^* estimates and their confidence limits to see how they differ. Also, the λ_k^* with the largest terms $(Y_k - \lambda^* t_k)^2/(\lambda^* t_k)$ of Q are the estimates that differ most significantly from λ^*.

Air conditioners. This test is used to compare Proschan's planes 7907 and 7908. The pooled estimate is $\lambda^* = 100(6 + 23)/(493 + 2201) = 0.01076$. The chi-square statistic is

$$Q = \left[\frac{(6 - 0.01076 \times 493)^2}{0.01076 \times 493} + \frac{(23 - 0.01076 \times 2201)^2}{0.01076 \times 2201}\right] = 0.11.$$

This statistic has $2 - 1 = 1$ degree of freedom. Since $Q = 0.11 < 3.841 = \chi^2(0.95; 1)$, the two planes' observed recurrence rates for repairs do not differ statistically significantly.

8.3 Nonhomogeneous Poisson Processes

Overview. This section briefly presents nonhomogeneous Poisson processes, which are simple models for counting processes for the cumulative number $Y(t)$ of recurrences over time t. They generalize the Poisson process. The section first defines such processes. It gives a nonparametric estimate and confidence limits for the MCF $M(t) \equiv EY(t)$. Next, it describes the most widely used such process, the power process. Then it briefly surveys the use of the power process to model reliability growth of products.

Definition. A counting process with cumulative integer values $Y(t)$, unit jumps, and independent increments is called a *nonhomogeneous Poisson process*. The MCF $M(t) = E[Y(t)]$ may be an arbitrary increasing function, usually assumed to be continuous and differentiable. Then at age t, the distribution of $Y(t)$ is Poisson with mean $M(t)$. In addition, for any interval (t_1, t_2), the distribution of the observed number $Y(t_1, t_2) = Y(t_2) - Y(t_1)$ of recurrences in that interval is Poisson with mean $M(t_1, t_2) = M(t_2) - M(t_1)$. That is,

$$\Pr\{Y(t_1, t_2) = y\} = \frac{1}{y!}[M(t_1, t_2)]^y \exp[-M(t_1, t_2)], \quad y = 0, 1, 2, 3, \ldots.$$

Figure 8.3 depicts a nonhomogeneous Poisson process, where $M(t)$ is the curve. The figure shows the Poisson distribution of the cumulative number of recurrences at two ages. This model is clearly a more flexible generalization of the simple Poisson process, which has $M(t) = \lambda t$ and is also called the *homogeneous Poisson process*. The property of independent increments is a mathematical convenience and often not valid in applications. Various functional forms for $M(t)$ have been used in applications. In particular, Rigdon and Basu (2000), Engelhardt (1995), and Lancaster (1990) present a variety of parametric MCFs. They provide estimates, confidence limits, and hypothesis tests for MCF parameters and $M(t)$. They and others denote the MCF $M(t)$ by $\Lambda(t)$ and the recurrence rate $m(t)$ by $\lambda(t)$.

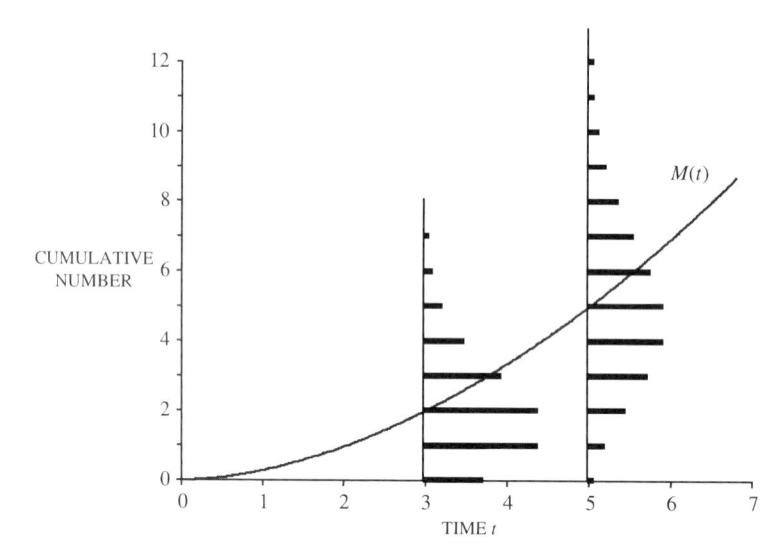

Figure 8.3. *Nonhomogeneous Poisson process depicted.*

Estimate and limits. Nonparametric estimates $M^*(t)$ for any $M(t)$ appear in previous chapters for exact and interval age data. The naive confidence limits for $M(t)$ are appropriate for a nonhomogeneous Poisson process. Vallarino (1988), for example, presents this estimate and naive confidence limits for exact age data and right censoring.

Naive limits. The following justifies the naive confidence limits (4.4) for $M(t)$ for a nonhomogeneous Poisson process using the estimate $M^*(t)$ from exact age data and right censoring. Suppose that the N sample units have censoring times $0 < \tau_N < \tau_{N-1} < \cdots < \tau_2 < \tau_1$, numbered backwards. The following variance $V[M^*(t)]$ is conditional on the censoring ages. Suppose that the corresponding observed numbers of recurrences in those

8.3. Nonhomogeneous Poisson Processes

intervals $(\tau_{i+1}, \tau_i]$ are $Y_N, Y_{N-1}, \ldots, Y_2, Y_1$. For an age t in interval I, denote the observed number of recurrences in the interval $(\tau_{I+1}, t]$ by $Y_I(t)$. Then the estimate $M^*(t)$ is

$$M^*(t) = \left[\frac{Y_N}{N}\right] + \left[\frac{Y_{N-1}}{N-1}\right] + \cdots + \left[\frac{Y_I(t)}{I}\right].$$

Because the process has independent increments, these terms are statistically independent. Thus the variance is

$$V[M^*(t)] = \frac{V[Y_N]}{N^2} + \frac{V[Y_{N-1}]}{(N-1)^2} + \cdots + \frac{V[Y_I(t)]}{I^2}.$$

For a nonhomogeneous Poisson process, each variance on the right is the expected number of recurrences times the number of units that went through that interval. That is, $V[Y_N] = N[M(\tau_N) - M(0)]$, $V[Y_{N-1}] = (N-1)[M(\tau_{N-1}) - M(\tau_N)], \ldots, V[Y_I(t)] = I[M(t) - M(\tau_{I+1})]$. Thus the true variance is

$$V[M^*(t)] = \frac{[M(\tau_N) - M(0)]}{N} + \frac{[M(\tau_{N-1}) - M(\tau_N)]}{N-1} + \cdots + \frac{[M(t) - M(\tau_{I+1})]}{I}.$$

Since Y_i/i is an unbiased estimate of $[M(\tau_i) - M(\tau_{i+1})]$, an unbiased estimate of $V[M^*(t)]$ is

$$v[M^*(t)] = \left[\frac{Y_N}{N^2}\right] + \left[\frac{Y_{N-1}}{(N-1)^2}\right] + \cdots + \left[\frac{Y_I(t)}{I^2}\right].$$

This is used to obtain the naive confidence limits (4.4). Section 4.3 applies this formula to obtaining naive confidence limits for interval age data.

Power process. The power process is widely used in applications. A nonhomogeneous Poisson process with the MCF

$$M(t) = \beta t^\alpha$$

is called the *power process*. Here the power α and multiplier β are positive parameters that must be estimated from data. Although known by the misnomer the *Weibull process*, it has little relation to a Weibull distribution. $M(t)$ here is called a *power relationship* or *power law*. "Law" here is a grandiose name for a simple curve with no theoretical basis and that is fitted to data. This MCF is a straight line on log-log paper. For example, Figure 3.4(b) lets one assess whether the power relationship is suitable for the compressor data. This model includes the simple Poisson model for $\alpha = 1$ and is an obvious generalization of it. The recurrence rate

$$m(t) = \alpha \beta t^{\alpha-1}$$

is also a power function of time. For $0 < \alpha < 1$, the rate decreases with time; for $\alpha = 1$, it is constant (Poisson); and for $\alpha > 1$, it increases with time.

Reliability growth. Used for reliability growth, the power process model is called the "Crow–AMSAA model." It is a management tool for monitoring reliability growth of an evolving fleet of units, as the product reliability is improved during development testing, manufacture, and use in customer service. Thus the model reflects ongoing improvements in design, manufacture, and operation. It is used for units that fail once and for repairable units. The data are summarized with the total number of failures $Y(T)$ of all units as a function of

T, the total running time summed over all units. The model for the expected value of the pooled failure rate is

$$E\left[\frac{Y(T)}{T}\right] = \beta' T^{\alpha-1}.$$

In practice, $\alpha - 1$ is negative when there is reliability growth (decreasing failure rate). Then $1 - \alpha$ is positive and called the growth rate, which is typically in the range 0.3–0.5. The basic technique developed by Duane (1964) involves a plot on log-log paper of $Y(T)$ versus T. This plot resembles yet differs from a sample MCF. The MCF for a stable population of units is a function of the age t of units. In contrast, here T is a sum of a mix of ages of an evolving population. If this "Duane plot" is straight, the power relationship is regarded as satisfactory and is extrapolated to predict future reliability. Crow (1982) developed statistical methods for the model, which also appear in IEC Standard 1164 (1995), Abernethy (2000, section 8.7), Tobias and Trindade (1995, Chapter 12), MIL-HDBK-781 (1987), and Rigdon and Basu (2000). These references present estimates and confidence limits for α, β', $M(t)$, and $m(t)$ and predictions and prediction limits for future numbers of failures. They also present more sophisticated models. The WinSMITH Visual package of Abernethy (2000) has features for Duane plotting and for fitting the power process model to data with confidence limits. ReliaSoft's (2000a, b) module RG has these and other features for reliability growth data.

8.4 Renewal Processes

Overview. This section briefly presents renewal processes. Renewal process theory appears in Tobias and Trindade (1995), Cox (1962), and Parzen (1999). This section

- defines a renewal process,

- gives the renewal equation for the relationship between the process MCF $M(t)$ and the cumulative life distribution $F(t)$ of times between replacements,

- solves that equation to obtain an estimate of $F(t)$ from the sample $M^*(t)$, and

- discusses preventive replacement, a special renewal process.

Definition. If each failed unit is immediately replaced with a new unit at that location, the number of replacements is a *renewal process* with an MCF $M(t)$, called the *renewal function*. This process assumes that the lifetimes of all units are statistically independent and from the same cumulative life distribution $F(t)$. Such replaced units are said to be "as good as new." The defrost control data (Problem 5.2) are renewal data, where one does not have the times between failures (unit life and censoring times); times between failures would be easy to analyze with life data methods. For a single braking grid location (Figure 7.3), the data are from a renewal process. Because each locomotive has six grids, a locomotive has six superimposed renewal processes, a more complicated process. In contrast, Kvam, Singh, and Whitaker (2002) and their references present repair models and data analyses where units are not restored to new condition.

Goal. In most applications, one uses recurrence data to estimate the population MCF, using the methods in previous chapters. Then one usually wants to estimate the population life distribution $F(t)$ from the sample MCF. A nonparametric estimate for $F(t)$ follows.

8.4. Renewal Processes

Renewal equation. The MCF $M(t)$ and cdf $F(t)$ are related through the renewal equation

$$M(t) = F(t) + \int_0^t M(t-x)\, dF(x).$$

This integral equation contains a convolution of $M(t)$ and $F(t)$. The Poisson process satisfies this equation, where $M(t) = \lambda t$ and $F(t) = 1 - \exp[-\lambda t]$. One can use a Laplace transformation of this equation to solve for $M(t)$ when $F(t)$ is given, or vice versa. When the life distribution is Weibull, the renewal function $M(t)$ cannot be expressed explicitly, but it has been numerically evaluated and tabulated by White (1964). The WeibullSmith software by Abernethy (2000) has features for Weibull analysis of renewal data.

$M(t)$ properties. For a renewal process, the renewal function $M(t)$ has the following properties. For small times t, $M(t) \cong F(t)$, say when both are less than 0.10 (10%), often the case in applications. For example, see Problem 8.4. For large times t (enough greater than the mean μ of the life distribution $F(t)$), $M(t) \sim M_0 + t/\mu$, where M_0 is a constant. That is, asymptotically the population recurrence rate is constant, and the MCF increases by 1 with each additional mean life μ, an intuitively obvious result.

Estimate $F(t)$. For interval age data, it is easy to estimate the cdf $F(t)$ from the sample MCF as follows. Suppose there are I intervals with endpoints $0 = \tau_0 < \tau_1 < \tau_2 < \cdots < \tau_I$, which need not be equally spaced. Denote the MCF estimates at these endpoints by $M_1, M_2, \ldots, M_I$. We seek the corresponding endpoint estimates $F_1, F_2, \ldots, F_I$ of the cumulative distribution. For each τ_i, treat the sample MCF as linear between successive M_i values. Then use this piecewise linear MCF in the renewal equation and approximate the convolution with a sum. This yields the following equations:

$$F_0 = M_0 = 0,$$

$$F_1 = \frac{M_1}{\left[1 + \frac{1}{2} M_1\right]},$$

$$F_2 = \frac{\left[M_2 - \frac{1}{2} M_2 F_1\right]}{\left[1 + \frac{1}{2} M_1\right]},$$

$$F_3 = \frac{\left\{\left[M_3 - \frac{1}{2} M_3 F_1\right] - \frac{1}{2} M_2 F_2 + \left[\frac{1}{2} M_1 F_1\right]\right\}}{\left[1 + \frac{1}{2} M_1\right]},$$

$$\vdots$$

$$F_i = \frac{\left\{\left[M_i - \frac{1}{2} M_i F_1\right] - \frac{1}{2} M_{i-1} F_2 - \frac{1}{2} M_{i-2}(F_3 - F_1) - \cdots - \frac{1}{2} M_3(F_{i-2} - F_{i-4}) - \frac{1}{2} M_2(F_{i-1} - F_{i-3}) + \left[\frac{1}{2} M_1 F_{i-2}\right]\right\}}{\left[1 + \frac{1}{2} M_1\right]},$$

$$\vdots$$

$$F_I = \frac{\left\{\left[M_I - \frac{1}{2} M_I F_1\right] - \frac{1}{2} M_{I-1} F_2 - \frac{1}{2} M_{I-2}(F_3 - F_1) - \cdots - \frac{1}{2} M_3(F_{I-2} - F_{I-4}) - \frac{1}{2} M_2(F_{I-1} - F_{I-3}) + \left[\frac{1}{2} M_1 F_{I-2}\right]\right\}}{\left[1 + \frac{1}{2} M_1\right]}.$$

This estimate for interval age data is like those presented by Trindade (1980), who also presents confidence limits for F_i. Tobias and Trindade (1995) present an estimate for exact age data. Trindade and Haugh (1990) use such estimates for the life distribution of circuit boards replaced in mainframe computers. Also, the equations above extend to exact age data.

As I gets large and equal-length intervals shrink to zero length, the equations in the limit yield an $F(t)$ estimate for exact age data. In practice, one uses enough intervals so that there are few intervals with more that one sample age. Peña, Strawderman, and Hollander (2001) survey literature on estimating $F(t)$ and give an estimate and corresponding approximate confidence limits.

Preventive replacement. A special renewal process arises where a unit is replaced at age τ if it has not failed by that age. Then τ is chosen to minimize the average cost per unit time of operation. A simple model involves the costs a and b of replacing a unit *after* and *before* it fails and the underlying life distribution. Glasser (1969) presents this cost model and optimizes τ for a Weibull life distribution. WinSMITH Visual of Abernethy (2000) has features for determining Glasser's optimal replacement age. Zaino (1987) and Zaino and Berke (1994) present applications. Jorgensen, McCall, and Radner (1967) treat replacement theory in more detail.

8.5 Models with Covariates

Overview. This section briefly surveys simple counting process models with covariates. Such models describe how the population MCF depends on covariates. This section first discusses covariates. Next, it presents the Poisson regression model with covariates. Then it presents the Cox regression model for recurrent events. Suitable software packages for fitting these models to data are referenced.

Covariates. In some applications, one has observed the values of K covariates $x_1, x_2, \ldots, x_K$ for each sample unit. Such covariates may be categorical (e.g., male and female or various treatments) or numerical (e.g., operating temperature, size, and years of schooling). Such covariates may be known functions (transformations) of more basic variables. Also, covariates may be time-varying, $x_1(t), x_2(t), \ldots, x_K(t)$, as discussed by Therneau and Grambsch (2000).

Poisson regression. The Poisson regression model expresses the recurrence rate λ as the following function of the covariates and coefficients $\beta_0, \beta_1, \ldots, \beta_K$:

$$\lambda(x_1, x_2, \ldots, x_K; \beta_0, \beta_1, \ldots, \beta_K) = \exp[\beta_0 + \beta_1 x_1 + \cdots + \beta_K x_K].$$

This log-linear relationship guarantees that λ is positive. Lawless (1987), Therneau and Grambsch (2000), Cameron and Trivedi (1998), and Winkelmann (2000) give methods for fitting this and other parametric regression models to count data. JMP and SAS software of the SAS Institute (1999, 2000), among others, fit such models to data. This simple model has a constant recurrence rate for any set of covariate values, which is unlikely in practice.

Cox model. The Cox regression relationship for recurrence data expresses the MCF as the following function of the covariates and coefficients:

$$M(t) = M_0(t) \exp[\beta_1 x_1 + \cdots + \beta_K x_K].$$

Here the nonparametric function $M_0(t)$ and the coefficients $\beta_1, \beta_2, \ldots, \beta_K$ are estimated from the data, as described by Therneau and Grambsch (2000), Lawless and Nadeau (1995), and Allison (1984, 1996). In the Poisson regression model, $M_0(t) = e^{\beta_0} t$ is specified. The Cox model implies that the MCF for any set of covariate values has the same shape and differs from MCFs for other covariate values only by the multiplier $\exp[\cdot]$. This may not hold in practice. For example, the men's and women's MCFs (Figure 5.1) clearly have different shapes. This Cox relationship is also used to model a cumulative hazard function $H(t)$ of a life distribution (section 2.3). Then the Cox relationship has an entirely different meaning and interpretation.

Software. Theory for fitting this model to recurrence data is simple, and many computer programs do such fitting. Thus the Cox model has been widely used despite its limitations. Such fitting employs the often dubious assumption of independent increments. Therneau and Grambsch (2000) show how to use statistical packages to fit this model to count data. Also, they show how to get suitable confidence limits that do not use the dubious assumption of independent increments. JMP and SAS software of the SAS Institute (1999, 2000) and S-PLUS of Insightful (2001), among others, do such fitting.

8.6 Other Models

Purpose. This section provides references on a number of other models that involve recurrence data.

Single unit. This book deals with data on a sample of units. In some applications, there is data from only one unit. Such data are typically analyzed by fitting a parametric model to the data. Key references are Ascher and Feingold (1984), Engelhardt (1995), and Rigdon and Basu (2000).

Markov processes. Markov chains and processes have been used to model many phenomena, especially those with a number of states. The Poisson process is a simple Markov process in which each number of recurrences is a state. Examples include states of system failure and repair, stages of a disease, and the changing employment status of people. Selected references follow:

- J. R. NORRIS (1998), *Markov Chains*, Cambridge University Press, Cambridge, UK.

- N. LIMNIOS AND G. OPRIŞAN (2001), *Semi-Markov Processes and Reliability*, Birkhäuser, Basel.

- I. N. KOVALENKO, N. Y. KUZNETZOV, AND P. A. PEGG (1997), *Mathematical Theory of Reliability of Time Dependent Systems with Practical Applications*, John Wiley, New York.

Warranty analysis. Warranty analysis is concerned with modeling and evaluating the costs of warranty policies. When the failed product is replaced with a new unit, renewal modeling is appropriate, and the MCF is the mean number of replacements per unit on warranty. Some warranty policies involve a prorated cost to the customer based on the "age" of the product on replacement, a common practice for car tires and batteries. The books by Blischke and Murthy (1994a, b) and their over 1000 references provide an introduction to this subject. Robinson and McDonald (1991), Lawless (1998), and Suzuki, Karim, and Wang (2001) discuss statistical methods for and practical problems with automotive warranty claim data.

Longitudinal data. Longitudinal data analysis concerns situations in which an observed variable $Y_i(t)$ of unit i is a function of time t. A simple example is the height of children as a function of age. The variable may be continuous (height), integer (counts of recurrences), or categorical (a criminal is in or out of prison, a machine is up or down). Then $Y_i(t)$ is observed (or sampled) at M_i discrete times $t_{i1}, t_{i2}, \ldots, t_{iM_i}$, which may be the same or different for all sample units. Some observations on a unit may be missing (censored), and such censoring may be noninformative or informative. Each $Y_i(t)$ is modeled with a linear or nonlinear relationship, whose coefficients are random, differing from unit to unit, plus an error term. Then the coefficients for a particular unit are an observation from a multivariate distribution of coefficients, for example, a multivariate normal. One then estimates the mean vector and sometimes the covariance matrix of that distribution of coefficients. The mean relationship (evaluated at the mean vector of coefficients) corresponds to the MCF in previous chapters. An application is growth models. Also, the $Y_i(t)$ may also be functions of K covariates $X_1, X_2, \ldots, X_K$. Books on this subject include Heckman and Singer (1985) and the following:

- M. DAVIDIAN AND D. M. GILTINAN (1995), *Nonlinear Models for Repeated Measurements Data*, Chapman and Hall, London.

- P. J. DIGGLE, K.-Y. LIANG, AND S. L. ZEGER (1994), *Analysis of Longitudinal Data*, Oxford University Press, New York.

- A. M. KSHIRSAGAR AND W. B. SMITH (1995), *Growth Curves*, Marcel Dekker, New York.

- J. K. LINDSEY (1993), *Models for Repeated Measurements*, Oxford University Press, New York.

- J. C. PINHIERO AND D. M. BATES (2000), *Mixed-Effects Models in S and S-PLUS*, Springer-Verlag, New York.

Problems

8.1. Proschan data. Use the Proschan data (Problem 2.2) on repairs of air conditioners in $K = 13$ planes. Proschan models the data with a separate Poisson process and λ_k for each plane.

(a) Calculate the recurrence rate estimate and corresponding 95% confidence limits for each plane.

(b) Calculate a pooled estimate of a common λ and corresponding 95% limits.

(c) Make a plot of the pooled estimate and the 13 separate estimates and corresponding confidence limits. Examine the plot and subjectively assess whether they differ convincingly. Describe what you see.

(d) Calculate the chi-square statistic to test for equality of the 13 rates. What do you conclude? How do the rates differ?

(e) In view of the four overhauls, would you recommend other analyses?

8.2. Tumor data. The MCF for the number of bladder tumors is a straight line for the placebo and Thiotepa treatments. Thus a simple Poisson model for each patient, each with a different recurrence rate, may be suitable. For the Thiotepa data (Problem 3.2), do the following Poisson analyses:

(a) Use the chi-squared test to assess whether the 38 patients can be modeled with a common recurrence rate λ.

(b) Suppose that the potential population of N patients has individual recurrence rates $\lambda_1, \lambda_2, \ldots, \lambda_N$. Denote the mean recurrence rate by $\lambda_0 = (\lambda_1 + \lambda_2 + \cdots + \lambda_N)/N$. Give a formula for an estimator for this mean rate.

(c) Show that your estimator is unbiased, and give formulas for its variance and standard error.

(d) Evaluate the estimate (b) and standard error (c) for the Thiotepa data.

(e) Use (d) to obtain approximate 95% confidence limits for λ_0.

(f) Repeat (a) and (d) for the placebo data (Table 1.2).

(g) Estimate the difference of the two mean recurrence rates.

(h) Calculate approximate 95% confidence limits for the difference.

8.3. Defrost control. Use the defrost control data of Problem 5.2, which are renewal data. Ignore the problems in the nonrandom sample.

(a) Estimate the population life distribution.

(b) Plot the estimate on Weibull probability paper. Comment on what the plot shows.

8.4. Gearboxes. Use the gearbox claim data in Problem 5.6. Gearboxes are replaced with new ones.

(a) Calculate and plot the sample MCF for gearboxes on cars sold in month 1. Plot the MCF on both a linear grid and a log-log grid.

(b) Use renewal theory to estimate the corresponding life distribution.

(c) Plot this distribution estimate on Weibull paper and interpret the plot. For example, as gearboxes age, are they more or less prone to a claim?

(d) Use the last column of data in Problem 5.6 to calculate a pooled sample MCF for all gearboxes. Plot this estimate on a linear grid and a log-log grid.

(e) Use renewal theory to estimate the corresponding life distribution.

(f) Plot this distribution estimate on Weibull paper and interpret the plot.

(g) Estimate the percentage of original gearboxes that will survive the 48-month warranty.

References

O. O. AALEN (1978), Nonparametric inference for a family of counting processes, *Ann. Math. Statist.*, 6, pp. 701–726.

R. ABERNETHY (2000), *The New Weibull Handbook*, 4th ed., Dr. Robert Abernethy, North Palm Beach, FL; available for purchase from weibull@worldnet.att.net.

R. AGRAWAL AND N. DOGANAKSOY (2001), Analysis of reliability data from in-house audit laboratory testing, in *Advances in Reliability*, N. Balakrishnan and C. R. Rao, eds., Handbook of Statistics 20, Elsevier Science, New York, Chapter 27, pp. 693–705.

P. D. ALLISON (1984), *Event History Analysis: Regression Analysis for Longitudinal Event Data*, Sage Publications, Newbury Park, CA.

P. D. ALLISON (1996), Fixed effects partial likelihood for repeated events, *Sociological Methods Res.*, 25, pp. 207–222.

K. ANDERSEN, O. BORGAN, R. D. GILL, AND N. KEIDING (1993), *Statistical Models Based on Counting Processes*, Springer-Verlag, New York.

D. F. ANDREWS AND A. M. HERZBERG (1985), *A Collection of Problems from Many Fields for the Student and Research Worker*, Springer-Verlag, New York.

H. ASCHER (2003), *Statistical Analysis of Systems Reliability*, Marcel Dekker, New York, to appear.

H. ASCHER AND H. FEINGOLD (1984), *Repairable Systems Reliability*, Marcel Dekker, New York.

R. D. BAKER (2001), Data-based modeling of the failure rate of repairable equipment, *Lifetime Data Anal.*, 7, pp. 65–83.

W. R. BLISCHKE AND D. N. P. MURTHY (1994a), *Warranty Cost Analysis*, Marcel Dekker, New York.

W. R. BLISCHKE AND D. N. P. MURTHY, EDS. (1994b), *Product Warranty Handbook*, Marcel Dekker, New York.

D. P. BYAR (1980), The Veterans Administration study of chemo prophylaxis for recurrent stage I bladder tumors, in *Bladder Tumors and Other Topics in Urological Oncology*, M. Pavone-Macaluso, P. H. Smith, and F. Edsmyn, eds., Plenum, New York, pp. 363–370.

A. C. CAMERON AND P. K. TRIVEDI (1998), *Regression Analysis of Count Data*, Cambridge University Press, Cambridge, UK.

M. R. Chernick (1999), *Bootstrap Methods: A Practitioner's Guide*, John Wiley, New York.

W. G. Cochran (1977), *Sampling Techniques*, 3rd ed., John Wiley, New York.

J. Cohen, D. Nagin, G. Wallstrom, and L. Wasserman (1998), Hierarchical Bayesian analysis of arrest rates, *J. Amer. Statist. Assoc.*, 93, pp. 1260–1270.

R. J. Cook, J. F. Lawless, and C. Nadeau (1996), Robust tests for treatment comparisons based on recurrent event responses, *Biometrics*, 52, pp. 557–571.

D. R. Cox (1962), *Renewal Theory*, John Wiley, New York.

L. H. Crow (1982), Confidence interval procedures for the Weibull process with applications to reliability growth, *Technometrics*, 24, pp. 67–72.

M. J. Crowder (2001), *Classical Competing Risks*, CRC Press, Boca Raton, FL.

H. A. David and M. L. Moeschberger (1978), *Theory of Competing Risks*, Griffin, London.

M. Davidian and D. M. Giltinan (1995), *Nonlinear Models for Repeated Measurements Data*, Chapman and Hall, London.

D. J. Davis (1952), An analysis of some failure data, *J. Amer. Statist. Assoc.*, 47, pp. 113–150.

A. C. Davison and D. V. Hinkley (1997), *Bootstrap Methods and Their Applications*, Cambridge University Press, Cambridge, UK.

P. J. Diggle, K.-Y. Liang, and S. L. Zeger (1994), *Analysis of Longitudinal Data*, Oxford University Press, New York.

N. Doganaksoy and W. Nelson (1991), *A Method and Computer Program MCFDIFF to Compare Two Samples of Repair Data*, TIS Report 91CRD172, Corporate Research and Development, General Electric Company, Schenectady, NY; also available from wnconsult@aol.com.

N. Doganaksoy and W. Nelson (1998), A method to compare two samples of recurrence data, *Lifetime Data Anal.*, 4, pp. 51–63.

J. T. Duane (1964), Learning curve approach to reliability monitoring, *IEEE Trans. Aerospace Electron. Systems*, 2, pp. 563–566.

B. Efron and R. J. Tibshirani (1994), *An Introduction to the Bootstrap*, CRC Press, Boca Raton, FL.

M. Engelhardt (1995), Models and analyses for the reliability of a single repairable system, in *Recent Advances in Life-Testing and Reliability*, N. Balakrishnan, ed., CRC Press, Boca Raton, FL, pp. 79–106.

T. R. Fleming and D. P. Harrington (1991), *Counting Processes and Survival Analysis*, John Wiley, New York.

G. J. Glasser (1969), Planned replacement: Some theory and its application, *J. Quality Tech.*, 1, pp. 110–119.

A. I. Goldman (1992), EVENTCHARTS: Visualizing survival and other timed-events data, *Amer. Statistician*, 46, pp. 13–18.

References

K. M. HARRIS (1996), Life after welfare: Women, work and repeat dependency, *Amer. Sociological Rev.*, 61, pp. 407–426.

J. J. HECKMAN AND B. SINGER, EDS. (1985), *Longitudinal Analysis of Labor Marker Data*, Cambridge University Press, Cambridge, UK.

X. J. HU, J. F. LAWLESS, AND K. SUZUKI (1998), Nonparametric estimation of a lifetime distribution when censoring times are missing, *Technometrics*, 40, pp. 3–13.

IEC STANDARD 1164 (1995), *Reliability Growth: Statistical Test and Estimation Methods*, L. H. Crow, principal author, International Electrotechnical Commission, Geneva.

INSIGHTFUL (2001), *S-PLUS Documentation*, Version 6.0, Insightful, Seattle, WA.

D. W. JORGENSON, J. J. MCCALL, AND R. RADNER (1967), *Optimal Replacement Policy*, Rand-McNally, Chicago.

J. D. KALBFLEISCH, J. F. LAWLESS, AND J. A. ROBINSON (1991), Methods for the analysis and prediction of warranty claims, *Technometrics*, 33, pp. 273–285.

I. N. KOVALENKO, N. Y. KUZNETZOV, AND P. A. PEGG (1997), *Mathematical Theory of Reliability of Time Dependent Systems with Practical Applications*, John Wiley, New York.

M. S. KRAATZ AND E. J. ZAJAC (1996), Exploring the limits of the new institutionalism: The causes and consequences of illegitimate organizational change, *Amer. Sociological Rev.*, 61, pp. 812–836.

A. M. KSHIRSAGAR AND W. B. SMITH (1995), *Growth Curves*, Marcel Dekker, New York.

P. H. KVAM, H. SINGH, AND L. R. WHITAKER (2002), Estimating distributions with increasing failure rate in an imperfect repair model, *Lifetime Data Anal.*, 8, pp. 53–67.

T. LANCASTER (1990), *Economic Analysis of Transition Data*, Cambridge University Press, London.

J. F. LAWLESS (1983), Statistical methods in reliability, *Technometrics*, 25, pp. 305–335.

J. F. LAWLESS (1987), Regression methods for Poisson process data, *J. Amer. Statist. Assoc.*, 82, pp. 808–815.

J. F. LAWLESS (1995a), The analysis of recurrent events for multiple subjects, *Appl. Statist.*, 44, pp. 487–498.

J. F. LAWLESS (1995b), Adjustments for reporting delays and the prediction of occurred but not reported events, *Canadian J. Statist.*, 22, pp. 15–31.

J. F. LAWLESS (1998), Statistical analysis of product warranty data, *Internat. Statist. Rev.*, 66, pp. 40–60.

J. F. LAWLESS AND J. C. NADEAU (1995), Some simple robust methods for the analysis of recurrent events, *Technometrics*, 37, pp. 158–168.

J. F. LAWLESS AND M. ZHAN (1998), Analysis of interval-grouped recurrent event data using piecewise-constant rate functions, *Canadian J. Statist.*, 26, pp. 549–565.

J. F. LAWLESS, X. J. HU, AND J. CAO (1995), Methods for the estimation of failure distributions and rates from automobile warranty data, *Lifetime Data Anal.*, 1, pp. 227–240.

J. J. LEE, K. R. HESS, AND J. A. DUBIN (2000), Extensions and applications of event charts, *Amer. Statistician*, 54, pp. 63–70.

N. LIMNIOS AND G. OPRIŞAN (2001), *Semi-Markov Processes and Reliability*, Birkhäuser, Basel.

J. K. LINDSEY (1993), *Models for Repeated Measurements*, Oxford University Press, New York.

M. D. MALTZ (1984), *Recidivism*, Academic Press, New York.

W. Q. MEEKER AND L. A. ESCOBAR (1998), *Statistical Methods for Reliability Data*, John Wiley, New York.

W. Q. MEEKER AND L. A. ESCOBAR (2002), *SPLIDA (S-PLUS Life Data Analysis) Software: Graphical User Interface*, Statistics Department, Iowa State University, Ames, IA; available online from www.public.iastate.edu/~splida.

MIL-HDBK-781 (1987), *Reliability Test Methods, Plans, and Environments for Engineering Development, Qualification and Production*, L. H. Crow, principal author, Naval Publications and Forms Center, Philadelphia, PA.

V. N. MORIN (2002), personal communication on studies of purchasing behavior of Internet customers, vmorin@amazon.com.

D. J. MYERS (1997), Racial rioting in the 1960s: An event history analysis of local conditions, *Amer. Sociological Rev.*, 62, pp. 94–112.

W. NELSON (1970), Confidence intervals for the ratio of two Poisson means and Poisson prediction intervals, *IEEE Trans. Reliability*, R-19, pp. 42–49.

W. NELSON (1979), *How to Analyze Data with Simple Plots*, ASQC Basic References in Quality Control: Statistical Techniques, Vol. 1, American Society for Quality, Milwaukee, WI.

W. NELSON (1982), *Applied Life Data Analysis*, John Wiley, New York.

W. NELSON (1988), Graphical analysis of system repair data, *J. Quality Tech.*, 20, pp. 24–35.

W. NELSON (1990), Hazard plotting of left truncated data, *J. Quality Tech.*, 22, pp. 230–238.

W. NELSON (1995a), Confidence limits for recurrence data: Applied to cost or number of repairs, *Technometrics*, 37, pp. 147–157.

W. NELSON (1995b), Analysis of repair data with two measures of usage, in *Recent Advances in Life-Testing and Reliability*, N. Balakrishnan, ed., CRC Press, Boca Raton, FL, pp. 51–57.

W. NELSON (1998), An application of graphical analysis of repair data, *Quality and Reliability Engrg. Internat.*, 14, pp. 49–52.

W. NELSON (2000a), Theory and applications of hazard plotting for censored failure data, reprinted in *Technometrics*, 42, pp. 12–25 as one of "Two Classics in Reliability Theory."

W. NELSON (2000b), Graphical comparison of sets of repair data, *Quality and Reliability Engrg. Internat.*, 16, pp. 235–241.

W. NELSON AND N. DOGANAKSOY (1989), *A Computer Program MCFDIFF for an Estimate and Confidence Limits for the Mean Cumulative Function for Cost or Number of Repairs of Repairable Products*, TIS Report 91CRD172, Corporate Research and Development, General Electric Company, Schenectady, NY; also available from wnconsult@aol.com.

E. T. M. NG AND R. J. COOK (1999), Robust inference for bivariate point processes, *Canadian J. Statist.*, 27, pp. 509–524.

J. R. NORRIS (1998), *Markov Chains*, Cambridge University Press, Cambridge, UK.

E. PARZEN (1999), *Stochastic Processes*, SIAM, Philadelphia.

E. A. PEÑA, R. L. STRAWDERMAN, AND M. HOLLANDER (2001), Nonparametric estimate with recurrent event data, *J. Amer. Statist. Assoc.*, 96, pp. 1299–1315.

J. C. PINHIERO AND D. M. BATES (2000), *Mixed-Effects Models in S and S-PLUS*, Springer-Verlag, New York.

F. PROSCHAN (2000), Theoretical explanation of observed decreasing failure rate, reprinted in *Technometrics*, 42, pp. 7–11 as one of "Two Classics in Reliability Theory."

RELIASOFT CORPORATION (2000a), *Weibull++6 Users Guide*, ReliaSoft Publishing, Tucson, AZ, www.reliasoft.com.

RELIASOFT CORPORATION (2000b), *Life Data Analysis Reference*, ReliaSoft Publishing, Tucson, AZ, www.reliasoft.com.

S. E. RIGDON AND A. P. BASU (2000), *Statistical Methods for the Reliability of Repairable Systems*, John Wiley, New York.

J. A. ROBINSON (1995), Standard errors for the mean cumulative number of repairs on systems from a finite population, in *Recent Advances in Life-Testing and Reliability*, N. Balakrishnan, ed., CRC Press, Boca Raton, FL, pp. 195–217.

J. A. ROBINSON AND G. C. MCDONALD (1991), Issues related to field reliability and warranty data, in *Data Quality Control: Theory and Pragmatics*, G. E. Liepins and V. R. R. Uppuluri, eds., Marcel Dekker, New York.

D. P. ROSS (1989), *Pilot Study of Commercial Water-Loop Heat Pump Compressor Life*, report under EPRI contract RP 2480-06, Electric Power Research Institute, Palo Alto, CA.

SAS INSTITUTE (1999), *SAS/QC® User's Guide*, Version 8, SAS Institute, Cary, NC.

SAS INSTITUTE (2000), *JMP Statistical Discovery Software: Statistics and Graphics Guide*, SAS Institute, Cary, NC, pp. 433–435.

K. SUZUKI (1985), Estimation of lifetime parameters from incomplete field data, *Technometrics*, 27, pp. 263–271.

K. SUZUKI (1993), Estimation of lifetime distribution using the relationship of calendar time and usage time, *Rep. Statist. Appl. Res.*, 40, pp. 10–22.

K. SUZUKI, M. R. KARIM, AND L. WANG (2001), Statistical analysis of reliability warranty data, in *Advances in Reliability*, N. Balakrishnan and C. R. Rao, eds., Handbook of Statistics 20, Elsevier Science, Amsterdam, Chapter 21, pp. 585–609.

P. F. THALL AND J. M. LACHIN (1988), Analysis of recurrent events: Nonparametric methods for random-interval count data, *J. Amer. Statist. Assoc.*, 83, pp. 339–347.

T. THERNEAU AND P. M. GRAMBSCH (2000), *Modeling Survival Data: Extending the Cox Model*, Springer-Verlag, New York.

T. THERNEAU AND S. A. HAMILTON (1997), rhDNase as an example of recurrent events analysis, *Statist. Medicine*, 16, pp. 2029–2047.

P. A. TOBIAS AND D. C. TRINDADE (1995), *Applied Reliability*, 2nd ed., CRC Press, Boca Raton, FL, Chapter 10 (Repairable systems I: Renewal processes), pp. 303–333, and Chapter 11 (Repairable systems II: Non-renewal processes), pp. 334–371.

D. C. TRINDADE (1980), *Nonparametric Estimation of a Lifetime Distribution Via the Renewal Function*, Ph.D. thesis, University of Vermont, Burlington, VT, dave@trindade.com.

D. C. TRINDADE AND L. D. HAUGH (1980), Estimation of the reliability of computer components from field renewal data, *Microelectron. Reliability*, 20, pp. 205–218.

C. R. VALLARINO (1988), *Confidence Bands for a Mean Value Function Estimated from a Sample of Right-Censored Poisson Processes*, presented at the 1988 Joint Statistical Meeting, New Orleans.

M. VASAN, D. PARKER, AND M. ALLEN (2000), A reliable model, *ActionLine*, September, pp. 36–40.

M.-C. WANG, J. QIN, AND C. T. CHIANG (2001), Analyzing recurrent events data with informative censoring, *J. Amer. Statist. Assoc.*, 96, pp. 1057–1065.

L. J. WEI, D. Y. LIN, AND L. WEISSFELD (1989), Regression analysis of multivariate incomplete failure-time data by modeling marginal distributions, *J. Amer. Statist. Assoc.*, 84, pp. 1065–1073.

J. S. WHITE (1964), Weibull renewal analysis, in *Proceedings of the Aerospace Reliability and Maintainability Conference*, American Society of Mechanical Engineers, New York, pp. 639–657.

R. WINKELMANN (2000), *Economic Analysis of Count Data*, Springer-Verlag, New York.

N. ZAINO (1987), Considerations in the use of optimal preventive maintenance models, *Quality and Reliability Engrg. Internat.*, 3, pp. 163–167.

N. ZAINO AND T. BERKE (1994), Some renewal theory results with application to fleet warranties, *Naval Res. Logist.*, 41, pp. 465–582.

Index

Actuarial adjustment, 82
Age
    censoring, 4
    definition, 4, 20
Age data
    distinct, 70
    exact, 3–11
    interval, 11–13
Analysis longitudinal, 133
Antibiotics, 58
Application, 1
    AIDS, 53
    air conditioner, 33, 76, 124–127, 134
    appliances, 2
    automotive, 1
    aviation, 2
    bladder tumor, 5–8, 55, 112, 134
    blood analyzer, 29
    braking grid, 112
    business, 2
    CGD, 58, 77, 118
    childbirth, 12, 31, 79, 81, 86, 90, 118
        to men, 80
        to women, 87
    circuit breaker, 58, 77
    compressor, 10, 46, 129
    computers, 1
    criminology, 2
    defrost control, 79, 87, 135
    economics, 2
    electric power, 1
    electronics, 1
    fan motor, 43, 66, 77
    fan motor costs, 8–10, 44
    gearbox, 91, 135
    gearbox costs, 93
    heat pump, 8, 10
    Internet customers, 10
    locomotives, 14, 56
    marketing, 2, 10
    medical equipment, 2
    medicine, 2
    military, 1
    power plant, 14
    Proschan, 33, 76, 124–127, 134
    refrigerators, 87
    reliability, 1
    social sciences, 2
    subway car, 16, 97
    traction motor, 16, 90, 97, 102, 103, 106, 108
    transmission, 4–5, 37, 55, 60, 76, 110
    transportation, 1
    trucks, 42
    tumor, 40, 61, 76, 90
    turbines, 18, 28
    valve seats, 56
Approximation
    χ^2, 116
    normal, 71
Assumptions
    confidence limits, 70
    confidence limits for a difference of two MCFs, 118
    dubious, 74
    MCF estimate, 51
    naive confidence limits, 86
    not assumed, 26, 74
    not used, 53
    unverified, 40
    verified, 51, 60
Autocorrelation, 71
Availability, 14, 30
    average, 31
    instantaneous, 31

Bad as old, 20
Bathtub, 29
Bias, 53

unknown, 36
Birth process, 122
Births, *see* Application, childbirth
Bladder tumors, *see* Application, bladder tumor
Blood analyzer, *see* Application, blood analyzer
Bonferroni inequality, 115
Bootstrap, 75
Burn-in, 29

Cause and effect, 108
Censoring, 4
    adjustment, 82
    age, 4, 69
    ages unknown, 42
    complex, 75
    conditional, 65
    estimate ages, 42
    informative, 25, 52, 76, 133
    interval age data, 82
    left, 10–11, 46–49
    left and gaps, 118
    noninformative, 52, 133
    nonrandom, 25, 52
        example, 52
    property of the data, 26, 99
    random, 52, 54, 70, 74
    right, 3–10, 37–43, 70
    times, 128
    unknown ages, 42
CGD, 58
Childbirth, *see* Application, childbirth
Chronic granulomatous disease, *see* Application, CGD
Coefficients, 132
    estimate, 132
    random, 134
Compare MCFs, 32, 109–119
    analysis-of-variance, 115
    bladder tumor, 112
    breaking grids, 112
    CGD, 118
    childbirth, 79, 118
    entire MCFs, 116–117
    entire range, 109
    issues, 118
    K samples, 117
    multivariate, 117
    naive, 119
    nonparametric, 116
    parametric, 116, 117
    plots, 110, 112, 116
    pointwise, 109
    Poisson rates, 127
    simple theory, 109
    theory, 118
    transmissions, 110
Competing risks, 108
Confidence bands, 63, 118
Confidence intervals
    simultaneous, 115
Confidence limits
    approximate, 63, 83
    assumptions, 68–76
    bootstrap, 75, 83
    conservative, 68
    correct for interval data, 83
    difference of two MCFs, 110
    interval age data, 83–87
    jackknife, 75
    MCF, 59–77
    motivation, 60
    naive, 83, 129 (*see also* Naive confidence limits)
    naive assumptions, 86
    naive for interval age data, 83, 85
    negative, 67, 75
    Nelson, 59–64, 75
    normal approximation, 60
    pointwise, 63, 110
    Poisson, 64
    Poisson χ^2, 125
    Poisson λ, 125
    Poisson approximation, 67
    Poisson normal, 125
    positive, 75
    properties, 61
    resampling, 75
    simultaneous, 63, 110
    staircase, 67
    steps for naive, 84
    theory, 68–76
    too narrow, 83
    two-sided, 60
Correlation, 71
    mix of events, 99
Corticosteroids, 58

Costs
    definition, 21
    fan motor, 8–10, 43
    gearbox, 93
    incremental, 69, 73
    negative, 10
    prorated, 133
    statistically independent, 74
    treatment, 31
    ultimate, 31
    warranty, 133
Counting process, 132
Counts
    sums of, 124
Covariance, 71
    matrix, 134
    population, 71, 73
Covariates, 132, 134
    time-varying, 132
Cow, 14
Cox model, 132
    estimate, 132
Cox regression relationship, 132
    for a cumulative hazard function, 132
    for an MCF, 132
Crack growth, 14
Crow–AMSAA model, 129
Cumulative distribution function, 28
Cumulative hazard function, 28
Cumulative history function, 5

Data (*see also* Applications)
    childbirth, 12
    cost, 10, 82
    errors, 21
    exact age, 35–58
    gap, 11
    grouped ages, 79
    inspection, 15
    interval age, 79–95
    interval tabulation, 83
    intervals differ, 80
    intervals in common, 80
    life, 100
    longitudinal, 133
    multivariate, 10
    peculiar, 35
    Proschan, 33, 55 (*see also* Application, Proschan)
    survival, 28
    value, 10
Defect growth, 14, 49
Dependent increments, 54
    example, 74
Design life, 5, 101, 104
Difference of MCF estimates, 110 (*see also* Compare MCFs)
    all pairwise, 115
Disease (*see also* Application, tumor and CGD)
    episodes, 5
Display, 9
    compressor, 10
    interval data, 12
    mix of events, 16
    time-event, 4, 16, 34
Distribution
    χ^2, 125
    continuous, 26
    density, 26
    discrete, 26
    exponential, 33
    F, 126
    joint, 99
    life, 28, 130
    multivariate, 134
    multivariate normal, 134
    normal, 71
    normal cdf, 124
    Poisson, 71
    population, 26
    renewal, 130
    sampling, 71
Downtime, 20

Error, 21
    measurement, 21
    reporting, 21, 108
Estimate
    actuarial, 82
    MCF, *see* MCF estimate
    Poisson λ, 124
    pooled, 134
Estimate MCF, *see* MCF estimate
Events (*see also* Recurrences)
    charts, 5
    combined, 97
    definition, 19

discrete, 3
eliminated, 16, 97
group of, 16, 97, 99
mix of types, 15–18, 97–108
separate, 97
statistically dependent, 108
statistically independent, 108
types, 15–18, 25
types combined, 103, 105
types definition, 98
value, 24
Exact ages, 3–11
Excel, 46, 49, 67, 86, 98

Failure
definition, 19
degradation, 19
intermittent, 19
unreported, 19
Failure modes
combined, 16, 97
dependent, 18
eliminated, 16, 97
group of, 16, 97
independent competing, 100
individual, 16
separate, 97
statistically dependent, 54
statistically independent, 54
Failure rate, 27 (*see also* Recurrence rate)
decreasing, 130
pooled, 130
Fleet, 53

Gaps, 10–11, 46–49
Good as new, 20
Group of events, 16
Grouping into age intervals, 12
Growth of defect, 49

Hazard function, 27, 28
cumulative, 28
History function, 5
continuous, 13, 24
for cost, 9
for cost of recurrences, 24
for number, 9
for number of recurrences, 23
independent, 54

mix of events, 98
population, 23–25
staircase, 26
terminated, 21, 25
vector, 98
Homogeneous Poisson process, *see* Poisson process
Hypothesis test
difference of two MCFs, 112

Increment
MCF, 38
observed, 81
Incremental costs
bivariate population, 74
sample, 74
Incremental mean number, 46
Incremental recurrence cost, 69
Increments
integer, 24
negative, 24
Independent increments, 33, 54, 74, 84, 86, 133
hypothesis test for, 74
nonhomogeneous Poisson process, 129
often not valid, 128
Infant mortality, 29, 40
Information sought, 28–33
Intensity function, 26 (*see also* Recurrence rate)
Interarrival times, 2
Poisson process, 124
Internet shopping, 32
Interval
endpoints, 83
last, 86
Interval age data, 79–95
common, 80
common set, 83
differing, 80
treated as exact, 41
Issues, practical and theoretical, 51–54, 107–108

Jackknife, 75
JMP, 7, 40, 64, 112, 132, 133

Laplace transformation, 131
Least squares, 40

Index

Left censoring, *see* Censoring, left
Like-new condition, 54
Like-old condition, 54
Limits
    confidence, *see* Confidence limits
    prediction, *see* Prediction limits
Longitudinal
    analysis, 133
    data, 133

Maintenance, 34
    planning, 126
    preventative replacement, 132
Marked point processes, 10
Markov
    chains, 133
    process, 133
MCF
    add types, 99
    confidence limits, 59–77
    definition, 26
    difference of two, 110
    finite, 53
    for a type of event, 99
    K samples, 115
    K-variate, 99
    linear, 42
    mean curve, 26
    nonhomogeneous Poisson process, 128
    nonparametric, 132
    parametric, 54
    percentage, 46
    Poisson process, 122
    power process, 129
    sample, 38 (*see also* MCF estimate)
    ultimate, 31, 80
MCF estimate, 28
    all events, 100–102
    assumptions, 51, 70
    assumptions interval data, 82
    at a specified age, 40
    biased, 52, 82
    common, 115
    comparisons, 109 (*see also* Compare MCFs)
    continuous history, 49–51
    curve, 40, 47, 81
    design life, 5
    difference of two, 110
    events combined, 103–105
    exact age data, 35–58
    for a group of events, 103–105
    for a single type of event, 102
    for cost, 43–46
    for number, 37–43
    for remaining types of events, 106
    for value, 43
    formula, 69
    gaps, 46–49
    interval age data, 79–83
    Lawless–Nadeau variance, 64
    left censoring, 46
    line segments, 81, 101
    motivation, 36
    naive variance estimate, 65
    nonparametric, 38, 40
    not assumed, 53
    percentage, 100
    plotted points, 82
    pooled, 91, 115
    sampling distribution, 70
    staircase, 40, 47
    steps, 38, 43, 46
    straight line segments, 80
    theoretical variance, 73
    theory, 51
    true naive variance, 65
    true variance, 60, 71
    unbiased, 40, 53
    variance, 70
    variance estimate, 60, 64, 73
    variance for interval data, 84
    warranty, 95
    with events eliminated, 105–107
MCFDIFF, 112
MCFLIM, 40, 64
Mean cost per unit, 43
Mean cumulative function, *see* MCF
Mean curve, 26, 51
    K-variate, 99
Mean number, 38
Mean time between failures, 56, 124
Mean time between recurrences, 42
Menopause, 81
MIL-HDBK-781, 130
Mileage, 4, 93
    customer, 4
    test, 4

versus days, 94
Mixture, 54
Model
    choice of, 121
    counting process, 121
    Cox, 132
    Crow–AMSAA, 129
    growth, 134
    Markov, 133
    mix of events, 98–100
    nonparametric, 25
    other, 133–134
    parametric, 35, 121, 133
    Poisson process, 122–124
    population, 25–28
    series-system, 108
    single unit, 133
    with covariates, 132–133
MTBF, 124

Naive confidence limits, 64–68
    calculation, 67
    for cost, 65
    for number, 64
    limitations, 64
    not for cost, 64
    steps, 67
Nonhomogeneous Poisson process, 65, 86, 127–130
    definition, 128
    depicted, 128
    estimate, 128
    MCF, 128
    naive confidence limits, 128
    parametric MCF, 128
    recurrence rate, 128
    variance, 129
Number at risk, 38, 43, 46
    approximate, 81
Number entering, 81

Overhaul, 33

Parametric models, 54
Patient losses, 53
Permutation tests, 115
Placebo treatment, 5, 55
Plot, 110
    accuracy, 42

    advantages, 35
    combined, 105
    comparison, 110, 112
    disadvantages, 35
    Duane, 130
    enhanced, 101
    hazard, 38
    linear grid, 47
    log-log, 43, 47, 87, 95, 129, 130
    point per interval, 82
    point per recurrence, 82
    reliability growth, 130
Poisson distribution, 122
    analysis, 124–126
    confidence limits for λ, 125
    cumulative distribution function (cdf), 124
    density function, 122
    estimate common λ, 126
    mean, 122
    multisample analyses, 126–127
    normal approximation, 124
    pooled λ estimate, 126
    pooled rate variance, 127
    prediction for a count, 125
    prediction limits, 126
    probability density, 122
    regression fitting, 132
    regression model, 132
    standard deviation, 124
    statistically independent counts, 126
    variance, 124
Poisson process, 54, 121–127
    confidence limits, 125
    definition, 122
    estimate, 124
    homogeneous, 128
    interarrival times, 124
    MCF, 122
    mixture, 33, 75
    model, 122–124
    nonhomogeneous, *see* Nonhomogeneous Poisson process
    pooled estimate, 125
    Proschan data, 33
    recurrence rate, 122, 123
        estimate λ, 124
    renewal, 131
    sample recurrence rate, 124

Index 149

single-sample analyses, 124–126
sum of counts, 124
Population, 19
    bivariate, 73
    cumulative history function, 23–25
    finite, 53
    infinite, 71
    mixture, 75
    model, 23–34
    stratified, 76
    target, 19, 26, 70
Power law, 129
Power process
    definition, 129
    MCF, 129
    Poisson process is a special case, 129
    recurrence rate, 129
Power relationship, 129
Practical issues, 18–21
Prediction, 31, 54, 125
    error, 125
    variance, 125
Prediction limits, 54, 91
    approximate, 126
    Poisson, 126
Probability density, 28
Process
    counting, 2, 128
    independent increments, 86
    power, *see* Power process
Proschan data, *see* Application, Proschan

$R/100$, 5
Random sampling
    simple, 70
Recurrence rate
    add types, 99
    bathtub, 29
    behavior, 28, 40
    constant, 27, 42, 83, 132
    decreasing, 27
    definition, 26
    estimate, 40
    for group of events, 99
    for types of events, 99
    increasing, 27
    nonhomogeneous Poisson process, 128
    Poisson process, 122
    power process, 129

Recurrences, *see* Events
References, 137–142
Regression
    Cox, 132
    parametric, 132
    Poisson, 132
Related topics, 121–135
Relationship
    log-linear, 132
Reliability
    demonstration, 97
    growth, 129, 130
        ReliaSoft, 130
        WinSMITH, 130
    growth confidence limits, 130
    growth prediction, 130
    growth prediction limits, 130
    growth rate, 130
    improvement, 129
    Procedure, 86
ReliaSoft, 40, 64, 86, 112, 130
Renewal process, 48, 54, 130–132
    cdf estimate, 131
    definition, 130
    equation, 131
    function, 130
    MCF properties, 131
    Poisson, 131
    recurrence rate, 131
    Weibull, 131
Repair
    definition, 19
Replacement, 19
    optimum age, 132
    preventive, 132
Replacement rates, 87
Reporting delays, 53, 91
Resampling, 75
Resampling limits, 75
Retirement, 32
    optimum age, 32
Right censoring, *see* Censoring, right

Sample, 36
    biased, 21, 52
    nonrandom, 19, 90
    of units, 133
    random, 19, 51
    self-selected, 52, 90

simple random, 51, 71
single unit, 133
Sample MCF, 38 (*see also* MCF estimate)
Sample size, 35
Sampling distribution, 70
    empirical, 75
    normal, 118
    test statistic, 117
SAS, 40–42, 61, 64, 81, 86, 98, 112, 132, 133
Seasonal effect, 53
Series system, 100, 108
Significant
    practically, 32
    statistically, 32
Single unit, 2
Size effect, 40
Software, 81
    continuous histories, 51
    cost data, 46
    Cox model, 133
    difference of two MCFs, 112
    exact age data, 40, 64
    interval age data, 86
    left censoring and gaps, 49
    mix of events, 98
    naive limits, 67, 86
    permutation tests, 115
    two-sample, 117
SPLIDA, 40, 56, 64, 112
S-PLUS, 40, 64, 112, 133
Spreadsheet, 46, 98
Spreadsheet calculation, 86
Staircase function, 8, 24, 67
    difference of two MCFs, 110
Standard error
    naive, 84
Statistic
    χ^2, 127
    quadratic, 117
    test, 112, 116, 119
    test equal Poisson rates, 127
    two-sample, 117
Statisticians
    childbirths, *see* Application, childbirth
Stochastic process, 2, 26 (*see also* Process)
    counting, 26
    parametric, 26
    renewal, 130–132
Stratification, 76
Subpopulations, 90
Survival data, 28

Termination, 20, 25
Test
    chi-squared, 134
    equal Poisson rates, 127
    equality of K MCFs, 116
    likelihood ratio, 116
    permutation, 115
    statistic, 116
Thiotepa treatment, 5, 55
Ties, 53
Time zero, 20
Times, 2
    between recurrences, 2, 124
    distributions, 2
    first recurrence, 2
    interarrival, 2, 124
    second recurrence, 2
Traction motor, *see* Application, traction motor
Transmission, *see* Application, transmission
    automatic, 4
    manual, 55
Truncation age, 4
Tumors, *see* Application, bladder tumor
Types of events, 15–18

Ultimate average number, 13
Uncertainty, 108
Underreporting, 91
Uptime, 14
Usage, 4
    mileage versus days, 94
    multiple, 20

Value data, 10 (*see also* Cost)
    negative, 63, 112
Variance
    conditional on censoring ages, 128
    contribution, 67
    population, 71
Variance estimate
    biased, 74

increments, 84
interval age data, 84
negative, 74
positive, 74
unbiased, 74

Warranty, 101, 104
    analysis, 133
    automotive, 133
    claims, 91
    costs, 91
    policies, 133
Wearout behavior, 29
Weibull process, 129
WeibullSmith, 131
Weibull++, 40, 64, 86, 112
Wholly significant difference, 115
WinSMITH, 130, 132